Handbook of Cathodic Corrosion Protection

Handbook of Cathodic Corrosion Protection

Editor

Rohit Verma

Handbook of Cathodic Corrosion Protection
Edited by **Rohit Verma**

Printed in 2017

ISBN: 978-1-68117-391-7

Library of Congress Control Number: 2015941580

© 2016 by

SCITUS Academics LLC,
616, Corporate Way, Suite 2, 4766,
Valley Cottage, NY 10989

www.scitusacademics.com

Notice

Contents

Preface

Corrosion is a naturally occurring cost, worth billions in the oil and gas sector. New regulations, stiffer penalties for non-compliance and aging assets are all leading companies to develop new technology, procedures and bigger budgets catering to one prevailing method of prevention, cathodic protection. Cathodic Corrosion Protection Systems: A Guide for Oil and Gas Industries trains on all the necessary reports, inspection criteria, corrective measures and critical standards needed on various oil and gas equipment, structures, tanks, and pipelines. Demands in the cathodic protection market have driven development for better devices and methods, helping to prolong the equipment and pipeline's life and integrity. Going beyond just looking for leaks, this handbook gives the engineer and manager all the necessary tools needed to put together a safe cathodic protection system, whether it is for buried casing while drilling, offshore structures or submarine pipelines.

Editor

Corrosion Behaviour of Magnesium Alloys Coated with Tin by Cathodic Arc Deposition in Nacl and Na$_2$SO$_4$ Solutions

Hikmet Altun and Hakan Sinici

Department of Mechanical Engineering, Ataturk University, 25240, Erzurum, Turkey

ABSTRACT

Magnesium-based light-metal alloys belong to a class of structural materials with increasing industrial attention. Magnesium alloys show the lowest density among the engineering metallic materials, low cost and large availability. However, the limitations according to mechanical strength and the low corrosion resistance restrict their practical application. In this study, TiN was coated on magnesium-based AZ91 magnesium–aluminium–zinc alloy using cathodic arc PVD process. The corrosion behaviours of uncoated and coated magnesium alloys in 1% NaCl, 3% NaCl and 3% Na_2SO_4 solutions and the influence of the coatings on the corrosion behaviour of the substrate were investigated utilizing potentiodynamic polarization tests. A potentiostat for electrochemical corrosion tests, a cathodic arc physical vapour deposition coating system for coating processes, a scanning electron microscopy for surface examination and elemental analysis of the coatings were used in this study. It was determined that corrosion resistance of magnesium alloys can be increased with TiN coating on the alloys using cathodic arc PVD process.

INTRODUCTION

The need for weight reduction, particularly in portable microelectronics, telecommunication, aerospace and automobile sectors has stimulated engineers to be more creative in their choice of materials. Magnesium and its alloys, with one quarter of the density of steel and only two-thirds that of aluminium, and a strength to weight ratio that far exceeds either, fulfils the role admirably, as an 'ultra light' alloy [1]. Magnesium is the lightest of the structural metals, which makes it one of the favoured materials to minimize vehicle weight and therefore to reduce exhaust gas emissions in transport applications [2]. Although Mg alloys have the highest strength-to-weight-ratio of all the structural metals, several drawbacks, above all the below average corrosion properties, restrict the application of unprotected magnesium alloys [3] and [4]. Their poor corrosion resistance has hindered its widespread use in many applications [5].

One of the most effective ways to prevent corrosion is to coat the base materials. Surface modification by coatings has become an essential step to improve the surface properties such as wear, corrosion and oxidation. Various conventional techniques are utilized for depositing the desired material on to the substrate to achieve the surface modification [6]. Coatings can protect a substrate by providing a barrier between the metal and its environment and/or through the presence of corrosion inhibiting chemicals in them [7].

It is very important that the protection of magnesium alloys from corrosion using surface treatments for their much more widespread usage. Many studies related to the surface modifications carried out on magnesium alloys and their protection ability against corrosion have been made in recent years. Due to process cleanliness and environmental comprehensively, research into physical vapour deposition (PVD) coatings has been an essential task to develop various advanced surface modification materials for industrial manufacturing [8]. There are a few studies [9], [10], [11], [12], [13], [14], [15], [16], [17] and [18] in the literature about the effect of PVD coatings on the corrosion behaviour of magnesium alloys, but the researches have increased more and more. At almost all of the studies, sputtering process has been used as PVD technique. In contrast to sputtering, cathodic arc deposition process is a recent developed PVD technology which has a rapider coating rate and also plays an important role in PVD coating methods [19]. In addition, the cathodic arc process for the deposition of hard coatings is well known for its high ionization efficiency in the plasma and allows the deposition of dense coatings [20]. To date, it has not been met any study related to the effect of cathodic arc PVD coatings on the corrosion resistance of magnesium alloys. In addition, nearly all of the mentioned above studies in the literature related to the effect of PVD coatings on the corrosion behaviour of magnesium alloys have been focused on the corrosion behaviours in solutions including Cl^- ions such as NaCl. However, the corrosion behaviour of magnesium alloys in the solutions including $(SO_4)^{-2}$ ions also is a serious attractive area [21]. But, it also has not been met any study about the effect of $(SO_4)^{-2}$ ions on the corrosion behaviour of the PVD coated magnesium alloys. Therefore, the purpose of this study is to deposit TiN coating on the magnesium alloys by cathodic arc deposition process, and then to evaluate the effect of the coating on the corrosion behaviour of the alloys in NaCl and Na_2SO_4 solutions.

EXPERIMENTAL DETAILS

In this study, AZ91 alloy which is one of the most commonly used magnesium alloys was used as the substrate material. The chemical composition of the alloy is reported in Table 1. Before coating, the surfaces of the samples were ground by SiC emery paper with grits of 400, 800 and 1200 and were polished by Al_2O_3 paste. Then, they were rinsed in distilled water and acetone, and dried in warm air. TiN coatings were deposited on the substrates by cathodic arc PVD technique according to the parameters listed in Table 2.

Table 1: The chemical composition of AZ91 magnesium alloy (weight %)

Al	Zn	Cu	Ni	Mn	Si
8.46	0.83	0.07	0.01	0.09	0.08

Table 2: The cathodic arc PVD parameters for TiN coating

Bias voltage	N_2 pressure	Cathode current	Coating time
– 200 V	4 mtorr	70 A	30 min

The electrochemical evaluation was carried out at room temperature using a standard three-electrode (reference, counter and working) configuration, with an Ag/AgCl electrode as the reference electrode, a platinum plate as the counter electrode and the specimen as the working electrode. The experiments were performed using a potentiostat controlled with a computer. The potentiodynamic polarization tests were conducted in 500 ml solutions of 1% NaCl, 3% NaCl and 3% Na_2SO_4 at a constant scan rate of 1 mV/s. Prior to polarization, the samples were allowed to stabilize to obtain a stable open circuit potential (OCP). A scanning electron microscopy (SEM) for surface examination and elemental analysis of the coatings were used.

RESULTS AND DISCUSSION

In Fig. 1a, SEM image of a small structural coating defect on TiN coated AZ91 magnesium alloy were given. As seen in the figure, the width of the defects could reach to about 25–30 µm. In addition, it is seen inFig. 1b that the micro cracks could be formed in the coating layer. These defects can form a direct path between a corrosive environment and the substrate and the risk of galvanic corrosion also exists. Hard material coatings are nobler than magnesium substrate, but small structural defects in the coating, e.g. pinholes, pores, or cracks, decrease the corrosion resistance of the composite [17]. These kinds of defects could arise from coating process or substrate. In spite of their excellent mechanical and tribological properties, their corrosion resistance has always been conditioned by the presence of structural defects such as pores, pinholes and cracks that appear during application [22]. A problem with cathodic arc vaporization source is that the arc causes the emission of molten globules (that is, macroparticles or macros) that deposit on the film surface. Thus, these coatings often exhibit porosity [23]. Every defect allowing corrosion medium to contact substrate surface leads to formation of a galvanic cell and pitting corrosion starts. The amount and the dimension of defects depend on the kind of PVD-process and on the deposition conditions, especially from the substrates temperature and the bias voltage [17]. If these defects are decreased, the protection ability of the coating from corrosion increases.

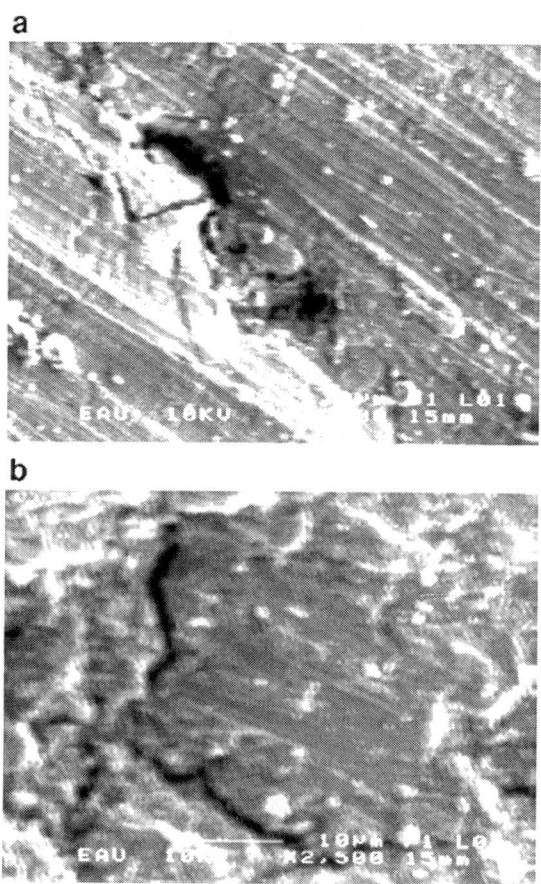

Figure 1: SEM images of small structural coating defects on TiN coated AZ91 magnesium alloy, a) a pinhole b) a crack.

Energy dispersive spectroscope (EDS) analysis of uncoated and coated alloys was carried out, and the graphics obtained were given in Fig. 2 and Fig. 3. EDS graphic of the uncoated magnesium alloy is seen in the Fig. 2. In the figure, the peaks of magnesium, aluminium and zinc elements were seen. EDS graphic of the TiN coated magnesium alloy is seen in the Fig. 3. As the figure was examined, differently from that of the uncoated status, it was obvious that the peaks of titanium element were formed.

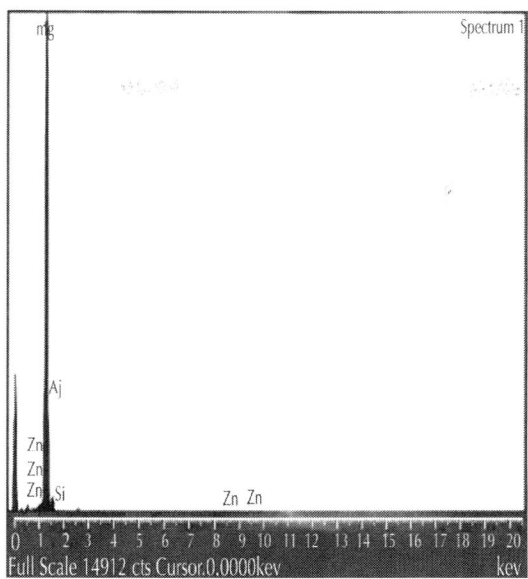

Figure 2: EDS graphic of uncoated AZ91 magnesium alloy.

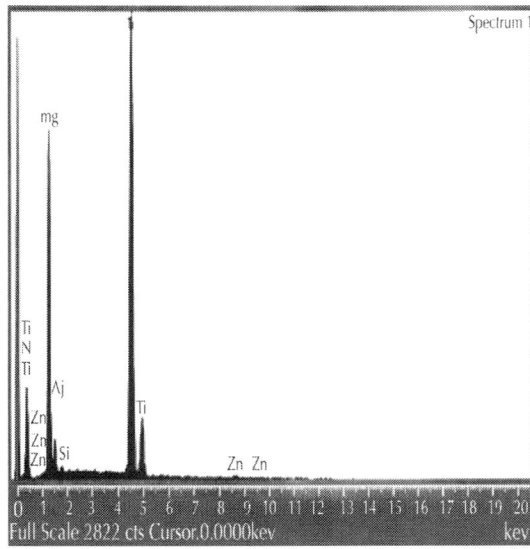

Figure 3: EDS graphic of TiN coated AZ91 magnesium alloy.

Fig. 4 and Fig. 5 show the potentiodynamic polarization curves of uncoated and TiN coated alloys in 1% and 3% NaCl solutions, respectively. As the figures were examined, it was seen that TiN coating decreased the anodic current densities of the alloy in that solutions, but the amount of the decreasing was small. In addition, it was determined that corrosion potentials of the uncoated and coated magnesium alloys were closed to each other, and no passivation behaviour was observed in 1% and 3% NaCl solutions. It could be said that TiN coating was not protective enough from corrosion in 1% and 3% NaCl solutions despite of a small amount decreasing at the anodic current densities obtained by means of TiN coating.

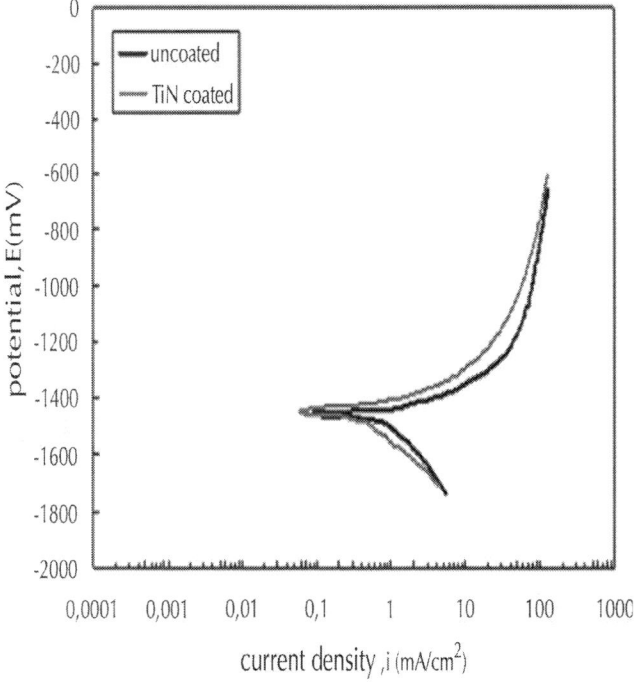

Figure 4: Potentiodynamic polarization curves of uncoated and TiN coated alloys in 1% NaCl solution.

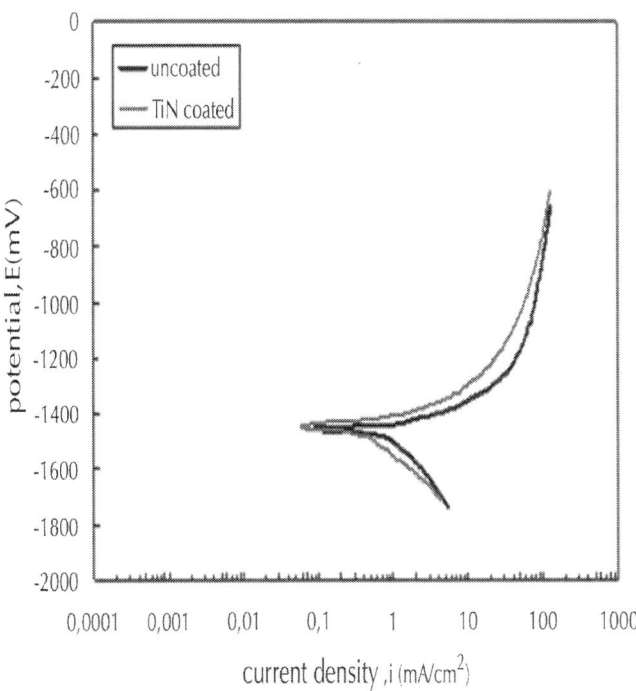

Figure 5: Potentiodynamic polarization curves of uncoated and TiN coated alloys in 3% NaCl solution.

Fig. 6 shows the potentiodynamic polarization curves of uncoated and TiN coated alloys in 3% Na_2SO_4. As the figure was examined, differently from that of 1% and 3% NaCl solutions, it was observed that TiN coating decreased the anodic current densities of the alloys at considerable amounts. Furthermore, it was obtained that corrosion current density was lower than that of the uncoated alloy, and corrosion potential value of TiN coated alloy was more positive than that of the uncoated alloy.

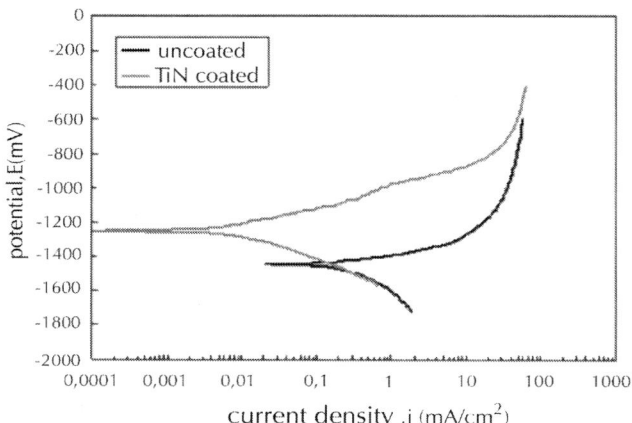

Figure 6: Potentiodynamic polarization curves of uncoated and TiN coated alloys in 3% Na$_2$SO$_4$ solution.

Potentiodynamic polarization curves obtained to determine the effect of solution exposed on the corrosion behaviour of the uncoated and coated magnesium alloys were given in Fig. 7 and Fig. 8. Fig. 7 shows the polarization curves of uncoated AZ91 magnesium alloy in 1% NaCl, 3% NaCl and 3% Na$_2$SO$_4$ solutions. As the anodic polarization curves were examined, it was seen that exposed solution affected anodic current densities at certain amounts. If an order according to corrosion resistance was done between those three solutions, it was seen that the most aggressive solution and the least aggressive solution for uncoated AZ91 magnesium alloy were 3% NaCl and 1% NaCl solutions, respectively. As the corrosion current densities were examined, it was seen that the highest corrosion current density was obtained in 3% NaCl solution, and the lowest corrosion current density was obtained in 1% NaCl solution.

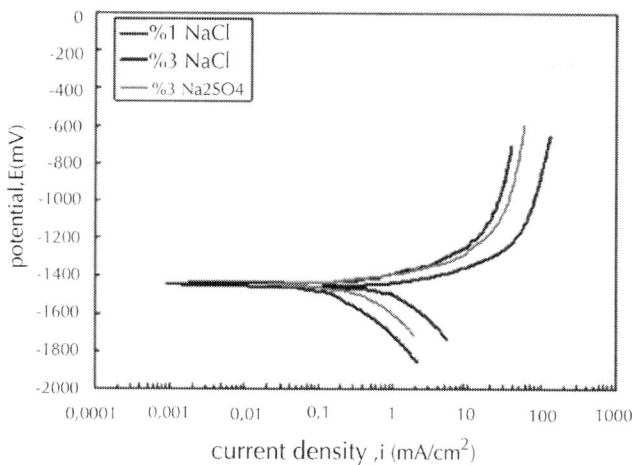

Figure 7: Potentiodynamic polarizaton curves of the uncoated magnesium alloys in 1% NaCl, 3% NaCl and 3% Na$_2$SO$_4$ solutions.

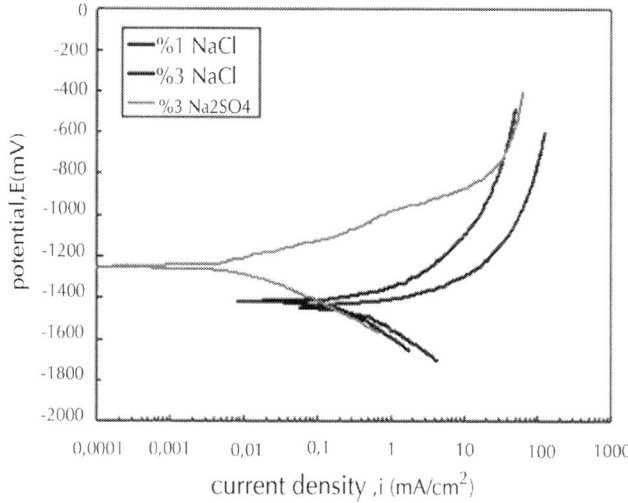

Figure 8: Potentiodynamic polarizaton curves of the TiN coated magnesium alloys in 1% NaCl, 3% NaCl and 3% Na$_2$SO$_4$ solutions.

Fig. 8 shows the polarization curves of TiN coated AZ91 magnesium alloy in 1% NaCl, 3% NaCl and 3% Na_2SO_4 solutions. As the anodic polarization curves were examined, it was observed that the anodic polarization curve of TiN coated AZ91 alloy in 3% Na_2SO_4 solution was at different character from that of the other solutions. The anodic current density of TiN coated alloy in 3% Na_2SO_4 solution was much lower than that of the other solutions until about − 800 mV potential. The current densities in the solution increased with increasing anodic polarization. As the corrosion current densities and corrosion potentials were examined, it was also seen that corrosion current density in 3% Na_2SO_4 solution was lower, and the corrosion potential was more positive in 3% Na_2SO_4 solution than that of the other solutions.

As mentioned above, corrosion behaviours of uncoated and TiN coated AZ91 alloy were investigated in three separate solutions. In NaCl solution, the tests were carried out at two separate concentrations (1% and 3%), and it was obtained that the corrosion resistance decreased for both uncoated and coated alloy with increasing in NaCl concentration. That is based on the increasing of the amount of chloride ion reacting for anodic dissolution and on the increasing of the conductivity of the solution with increasing in NaCl concentration. In addition, for inspection the aggressiveness of chloride and sulfate ions, the corrosion tests were carried out in 3% NaCl and 3% Na_2SO_4, and it was obtained that the corrosion resistance for both uncoated and coated alloy was lower in NaCl solution than that of Na_2SO_4 solution. It could be resulted from that chloride ions are more aggressive than sulfate ions for magnesium alloys. It can be said that the presence of chloride ions in the solution affects the passivity of magnesium alloy substrate more negatively than that of sulfate ions.

Surface images of TiN coated alloys after corrosion tests were given in Fig. 9. When the SEM images were examined, it was seen that the coating layer was broken from the substrate as large pieces, causing removal of the larger pieces. The formation of the coating defects is very much difficult to avoid totally. Consequently, when subjected to a corrosive atmosphere, coated materials will form galvanic cells at the defects near the interface since ceramic coatings are electrochemically more stable than most substrate materials. Once aggressive ions such as chloride penetrate the coating through these small channels, driven by capillary forces, the exposed area will begin to experience anodic dissolution, which will usually extend laterally along the interface between the coating and the substrate. Finally the pits to be formed linked up each other, causing removal of the entire coating by flaking [24]. The coating defects e.g., pores, pinholes, cracks, observed at the SEM images before corrosion tests lead the electrolyte to reach to the substrate and contribute to developing of the corrosion.

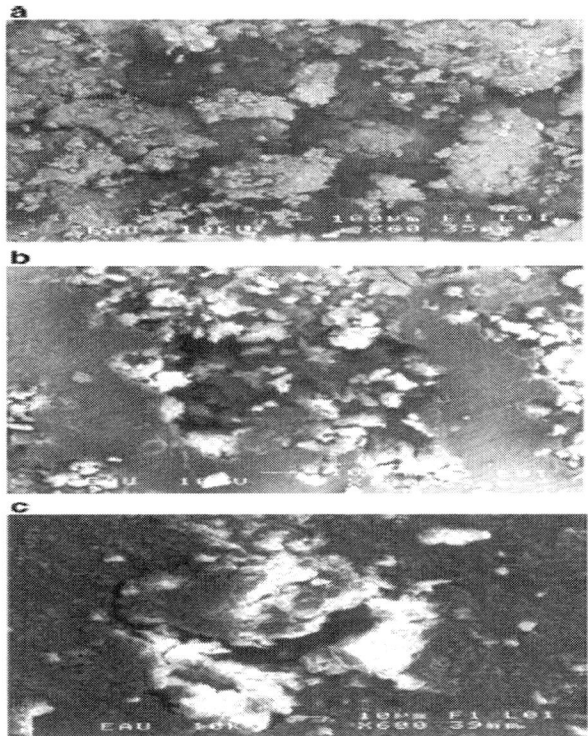

Figure 9: Surface images of TiN coated alloys after corrosion tests in a) 1% NaCl, b) 3% NaCl and c) 3% Na$_2$SO$_4$ solutions.

CONCLUSIONS

- TiN coating by cathodic arc deposition improved the corrosion resistance of the magnesium alloy. Especially in Na_2SO_4 solution, a considerable increasing in corrosion resistance was obtained.
- The coating layer had some typical defects resulting from PVD technique although relatively dense coatings were obtained.
- The corrosion resistance decreased for both uncoated and coated alloy with increasing in NaCl concentration.
- The corrosion resistance for both uncoated and coated alloy was higher in Na_2SO_4 solution than that of NaCl solution.

ACKNOWLEDGMENT

This work was funded by the Ataturk University Research Foundation.

REFERENCES

1.   Wu G, Fan Y, Gao H, Zhai C, Zhu YP. The effect of Ca and rare earth elements on the microstructure, mechanical properties and corrosion behavior of AZ91D. Mater Sci Eng A Struct Mater Prop Microstruct Process 2005;408:255–63.

2.   Boinet M, Verdier S, Maximovitch S, Dalard F. Plasma electrolytic oxidation of AM60 magnesium alloy: monitoring by acoustic emission technique, electrochemical properties of coatings. Surf Coat Technol 2005;199:141–9.

3.   Song G, Atrens A. Understanding magnesium corrosion — a framework for improved alloy performance. Adv Eng Mater 2003;5:837–58.

4.   Yamamoto A, Watanabe A, Sugahara K, Tsubakino H, Fukumoto S. Improvement of corrosion resistance of magnesium alloys by vapor deposition. Scr Mater 2001;44:1039–42.

5. Guo H, An M, Xu S, Huo H. Microarc oxidation of corrosion resistant ceramic coating on a magnesium alloy. Mater Lett 2006;60:1538–41.

6. Jagielski J, Khanna AS, Kucinski J, Mishra DS, Racolta P, Sioshansi P, et al. Effect of chromium nitride coating on the corrosion and wear resistance of stainless steel. Appl Surf Sci 2000;156:47–64.

7. Gray JE, Luan B. Protective coatings on magnesium and its alloys — a critical review. J Alloys Compd 2002;336:88–113.

8. Ho WY, Hsu CH, Huang DH, Lin YC, Chang CL, Wang DY. Corrosion behaviors of Cr(N,O)/CrN double-layered coatings by cathodic arc deposition. Surf Coat Technol 2005;200:1303–9.

9. Wu G, Zeng X, Ding W, Guo X, Yao S. Characterization of ceramic PVD thin films on AZ31 magnesium alloys. Appl Surf Sci 2006;252:7422–9.

10. Altun H, Sen S. The effect of PVD coatings on the corrosion behaviour of AZ91 magnesium alloy. Mater Des 2006;27:1174–9.

11. Altun H, Sen S. The effect of DC magnetron sputtering AlN coatings on the corrosion behaviour of magnesium alloys. Surf Coat Technol 2005;197:193–200.

12. Hoche H, Rosenkranz C, Delp A, Lohrengel MM, Broszeit E, Berger C. Investigation of the macroscopic and microscopic electrochemical corrosion behaviour of PVD-coated magnesium die cast alloy AZ91. Surf Coat Technol 2005;193:178–84.

13. Hoche H, Blawert C, Broszeit E, Berger C. Galvanic corrosion properties of differently PVD-treated magnesium die cast alloy AZ91. Surf Coat Technol 2005;193:223–9.

14. Hoche H, Scheerer H, Probst D, Broszeit E, Berger C. Plasma anodisation as an environmental harmless method for the corrosion protection of magnesium alloys. Surf Coat Technol 2003;174–175:1002–7.

15. Hoche H, Scheerer H, Probst D, Broszeit E, Berger C.

Development of a plasma surface treatment for magnesium alloys to ensure sufficient wear and corrosion resistance. Surf Coat Technol 2003;174–175:1018–23.

16. Lee MH, Bae IY, Kim KJ, Moon KM, Oki T. Formation mechanism of new corrosion resistance magnesium thin films by PVD method. Surf Coat Technol, 2003;169–170:670–4.

17. Hollstein F, Wiedemann R, Scholz J. Characteristics of PVDcoatings on AZ31hp magnesium alloys. Surf Coat Technol 2003;162:261–8.

18. Reiners G, Griepentrog M. Hard coatings on magnesium alloys by sputter deposition using a pulsed d.c. bias voltage. Surf Coat Technol 1995;76–77:809–14.

19. Cheng HH, Ming LC, Kuei LL. Corrosion resistance of TiN/TiAlN-coated ADI by cathodic arc deposition. Mater Sci Eng A 2006;421:182–90.

20. Chang YY, Wang DY. Characterization of nanocrystalline AlTiN coatings synthesized by a cathodic-arc deposition process, Surf Coat Technol in press. Corrected Proof, doi:10.1016/j.surfcoat.2006.09.0362.

21. Makar GL, Kruger J. Corrosion of magnesium. Int Mater Rev 1993;38:138–53.

22. Conde A, Navas C, Cristóbal AB, Housden J, Damborenea J de. Characterisation of corrosion and wear behaviour of nanoscaled e-beam PVD CrN coatings. Surf Coat Technol 2006;201:2690–5.

23. Ahn SH, Yoo JH, Choi YS, Kim JG, Han JG. Corrosion behavior of PVD-grown WC–(Til–xAlx)N films in a 3.5% NaCl solution. Surf Coat Technol 2003;162:212–21.

24. Dong H, Sun Y, Bell T. Enhanced corrosion resistance of duplex coatings. Surf Coat Technol 1997;90:91–101.

Chapter 2

Green Inhibitors for Corrosion Protection of Metals and Alloys: An Overview

B. E. Amitha Rani and Bharathi Bai J. Basu

Surface Engineering Division, CSIR-National Aerospace Laboratories, Bangalore 560037, India

ABSTRACT

Corrosion control of metals is of technical, economic, environmental, and aesthetical importance. The use of inhibitors is one of the best options of protecting metals and alloys against corrosion. The environmental toxicity of organic corrosion inhibitors has prompted the search for green corrosion inhibitors as they are biodegradable, do not contain heavy metals or other toxic compounds. As in addition to being environmentally friendly and ecologically acceptable, plant products are inexpensive, readily available and renewable. Investigations of corrosion inhibiting abilities of tannins, alkaloids, organic,amino acids, and organic dyes of plant origin are of interest. In recent years, sol-gel coatings doped with inhibitors show real promise. Although substantial research has been devoted to corrosion inhibition by plant extracts, reports on the detailed mechanisms of the adsorption process and identification of the active ingredient are still scarce. Development of computational modeling backed by wet experimental results would help to fill this void and help understand the mechanism of inhibitor action, their adsorption patterns, the inhibitor-metal surface interface and aid the development of designer inhibitors with an understanding of the time required for the release of self-healing inhibitors. The present paper consciously restricts itself mainly to plant materials as green corrosion inhibitors.

INTRODUCTION

Corrosion is the deterioration of metal by chemical attack or reaction with its environment. It is a constant and continuous problem, often difficult to eliminate completely. Prevention would be more practical and achievable than complete elimination. Corrosion processes develop fast after disruption of the protective barrier and are accompanied by a number of reactions that change the composition and properties of both the metal surface and the local environment, for example, formation of oxides, and diffusion of metal cations into the coating matrix, local pH changes, and electrochemical potential. The study of corrosion of mild steel and iron is a matter of tremendous theoretical and practical concern and as such has received a considerable amount of interest. Acid solutions, widely used in industrial acid cleaning, acid descaling, acid pickling, and oil well acidizing, require the use of corrosion inhibitors in order to restrain their corrosion attack on metallic materials.

CORROSION INHIBITORS

Over the years, considerable efforts have been deployed to find suitable corrosion inhibitors of organic origin in various corrosive media [1–4]. In acid media, nitrogen-base materials and their derivatives, sulphur-containing compounds, aldehydes, thioaldehydes, acetylenic compounds, and various alkaloids, for example, papaverine, strychnine, quinine, and nicotine are used as inhibitors. In neutral media, benzoate, nitrite, chromate, and phosphate act as good inhibitors. Inhibitors decrease or prevent the reaction of the metal with the media. They reduce the corrosion rate by

- adsorption of ions/molecules onto metal surface,
- increasing or decreasing the anodic and/or cathodic reaction,
- decreasing the diffusion rate for reactants to the surface of the metal,
- decreasing the electrical resistance of the metal surface.
- inhibitors that are often easy to apply and have in situ application advantage.

Several factors including cost and amount, easy availability and most important safety to environment and its species need to be considered when choosing an inhibitor.

Organic Inhibitors

Organic inhibitors generally have heteroatoms. O, N, and S are found to have higher basicity and electron density and thus act as corrosion inhibitor. O, N, and S are the active centers for the process of adsorption on the metal surface. The inhibition efficiency should follow the sequence $O < N < S < P$. The use of organic compounds containing oxygen, sulphur, and especially nitrogen to reduce corrosion attack on steel has been studied in some detail. The existing data show that most organic inhibitors adsorbed on the metal surface by displacing water molecules on the surface and forming a compact barrier. Availability of nonbonded (lone pair) and p-electrons in inhibitor molecules facilitate electron transfer from the inhibitor to the metal. A coordinate covalent bond involving transfer of electrons from inhibitor to the metal surface may be formed. The strength of the chemisorption bond depends upon the electron density on the donor atom of the functional group and also the polarizability of the group. When an H atom attached to the C in the ring is replaced by a substituent group ($-NH_2$, $-NO_2$, $-CHO$, or $-COOH$) it improves inhibition [4]. The electron density in the metal at the point of attachment changes resulting in the retardation of the cathodic or anodic reactions. Electrons are consumed at the cathode and are furnished at the anode. Thus, corrosion is retarded. Straight chain amines containing between three and fourteen carbons have been examined. Inhibition increases with carbon number in the chain to about 10 carbons, but, with higher members, little increase or decrease in the ability to inhibit corrosion occurs. This is attributed to the decreasing solubility in aqueous solution with increasing length of the hydrocarbon chain. However, the presence of a hydrophilic functional group in the molecule would increase the solubility of the inhibitors.

The performance of an organic inhibitor is related to the chemical structure and physicochemical properties of the compound like functional groups, electron density at the donor atom, p-orbital character, and the electronic structure of the molecule. The inhibition could be due to (i) Adsorption of the molecules or its ions on anodic and/or cathodic sites, (ii) increase in cathodic and/or anodic over voltage, and (iii) the formation of a protective barrier film. Some factors that contribute to the action of inhibitors are

- chain length,
- size of the molecule,
- bonding, aromatic/conjugate,
- strength of bonding to the substrate,
- cross-linking ability,
- solubility in the environment.

The role of inhibitors is to form a barrier of one or several molecular layers against acid attack. This protective action is often associated with chemical and/or physical adsorption involving a variation in the charge of the adsorbed substance and transfer of charge from one phase to the other. Sulphur and/or nitrogen-containing heterocyclic compounds with various substituents are considered to be effective corrosion inhibitors. Thiophene, hydrazine derivatives offer special affinity to inhibit corrosion of metals in acid solutions. Inorganic substances such as phosphates, chromates, dichromates, silicates, borates, tungstates, molybdates, and arsenates have been found effective as inhibitors of metal corrosion. Pyrrole and derivatives are believed to exhibit good protection against corrosion in acidic media. These inhibitors have also found useful application in the formulation of primers and anticorrosive coatings, but a major disadvantage is their toxicity and as such their use has come under severe criticism. Among the alternative corrosion inhibitors, organic substances containing polar functions with nitrogen, sulphur, and/or oxygen in the conjugated system have been reported to exhibit good inhibiting properties. The inhibitive characteristics of such compounds derive from the adsorption ability of their molecules, with the polar group acting as the reaction center for the adsorption process. The resulting adsorbed film acts as a barrier that separates the metal from the corrodent, and efficiency of inhibition depends on the mechanical, structural, and chemical characteristics of the adsorption layers formed under particular conditions.

Inhibitors are often added in industrial processes to secure metal dissolution from acid solutions. Standard anti corrosion coatings developed till date passively prevent the interaction of corrosion species and the metal. The known hazardous effects of most synthetic organic inhibitors and the need to develop cheap, nontoxic and ecofriendly processes have now urged researchers to focus on the use of natural products. Increasingly, there is a need to develop sophisticated new generation coatings for improved performance, especially in view of Cr VI being banned and labeled as a carcinogen. The use of inhibitors is one of the best options of protecting metals against corrosion. Several inhibitors in use are either synthesized from cheap raw material or chosen from compounds having heteroatoms in their aromatic or long-chain carbon system. However, most of these inhibitors are toxic to the environment. This has prompted the search for green corrosion inhibitors.

GREEN INHIBITORS

Green corrosion inhibitors are biodegradable and do not contain heavy metals or other toxic compounds. Some research groups have reported the successful use of naturally occurring substances to inhibit the corrosion of metals in acidic and alkaline environment. Delonix regia extracts inhibited the corrosion of aluminum in hydrochloric acid solutions [5], rosemary leaves were studied as corrosion inhibitor for the Al + 2.5Mg alloy in a 3% NaCl solution at 25°C [6], and El-Etre investigated natural honey as a corrosion inhibitor for copper [7] and investigated opuntia extract on aluminum [8]. The inhibitive effect of the extract of khillah (Ammi visnaga) seeds on the corrosion of SX 316 steel in HCl solution was determined using weight loss measurements as well as potentiostatic technique. The mechanism of action is attributed to the formation of insoluble complexes as a result of interaction between iron cations, and khellin [9] and Ebenso et al. showed the inhibition of corrosion with ethanolic extract of African bush pepper (Piper guinensis) on mild steel [10];Carica papaya leaves extract [11]; neem leaves extract (Azadirachta indica) on mild steel in H_2SO_4 [12]. Zucchi and Omar investigated plant extracts of Papaia, Poinciana pulcherrima, Cassia occidentalis, and Datura stramonium seeds and Papaya, Calotropis procera B, Azadirachta indica, and Auforpio turkiale sap for their corrosion inhibition potential and found that all extracts except those of Auforpio turkiale and Azadirachta indica reduced the corrosion of steel with an efficiency of 88%–96% in 1 N HCl and with a slightly lower efficiency in 2 N HCl. They attributed the effect to the products of the hydrolysis of the protein content of these plants [13]; Umoren et al. [14], studied the corrosion inhibition of mild steel in H_2SO_4 in the presence of gum arabic (GA) (naturally occurring polymer) and polyethylene glycol (PEG) (synthetic polymer). It was found that PEG was more effective than gum arabic.

Yee [15] determined the inhibitive effects of organic compounds, namely, honey and Rosmarinus officinalis L on four different metals—aluminium, copper, iron, and zinc, each polarized in two different solutions, that is, sodium chloride and sodium sulphate. The experimental approach employed potentiodynamic polarization method. The best inhibitive effect was obtained when zinc was polarised in both honey-added sodium chloride and sodium sulphate solutions. Rosemary extracts showed some cathodic inhibition when the metal was polarized in sodium chloride solution. This organic compound, however, displayed less anodic inhibition when compared with honey. The main chemical components of rosemary include borneol, bornyl acetate, camphor, cineole, camphene, and alpha-pinene. Chalchat et al. [16], reported that oils of rosemary were found to be rich in 1,8-cineole, camphor, bornyl acetate, and high amount of hydrocarbons. Recently, work has been emphasized on the use of Rosmarinus officinalis L as corrosion inhibitor for Al-Mg corrosion in chloride solution [6]. It is believed that the catechin fraction present in the rosemary extracts contributes to the inhibitive properties that act upon the alloy. Ouariachi et al. [17] also reported the inhibitory action ofRosmarinus officinalis oil as green corrosion inhibitors on C38 steel in 0.5 M H_2SO_4.

Odiongenyi et al. [18] reported that the ethanolic extract of Vernonia amygdalina appears to be a good inhibitor for the corrosion of mild steel in H_2SO_4 and action is by classical Langmuir adsorption isotherm.

The effect of addition of halides (KCl, KBr, and KI) was also studied, and the results obtained indicated that the increase in efficiency was due to synergism [13]. Umoren et al. also investigated the corrosion properties ofRaphia hookeri exudates gum—halide mixtures for aluminum corrosion in acidic medium [16]. Raphia hookeri exudates gum obeys Freundlich, Langmuir, and Temkin adsorption isotherms. Phenomenon of physical adsorption is proposed. Abdallah also tested the effect of guar gum on carbon steel. It is proposed that it acts as a mixed type inhibitor [14]. The mechanism of action of C-steel by guar gum is due to the adsorption at the electrode/solution interface. Guar gum is a polysaccharide compound containing repeated heterocyclic pyrane moiety as shown in Scheme 1. The presence of heterooxygen atom in the structure makes possible its adsorption by coordinate type linkage through the transfer of lone pairs of electron of oxygen atoms to the steel surface, giving a stable chelate five-membered ring with ferrous ions. The chelation between O1 and O2 with Fe++ seems to be impossible due to proximity factor presented as in Scheme 1:

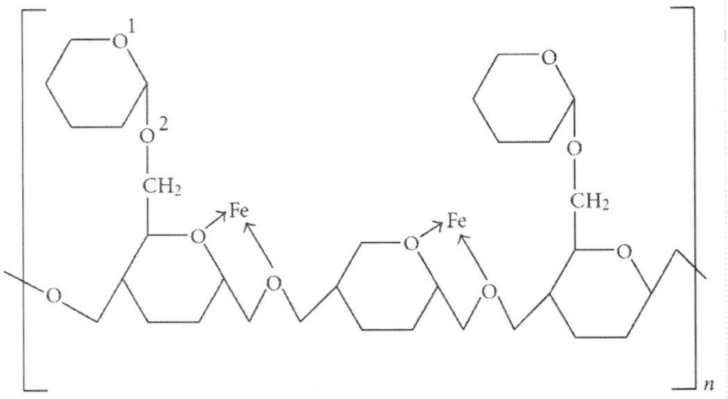

Scheme 1: Guar gum.

The simultaneous adsorption of oxygen atoms forces the guar gum molecule to be horizontally oriented at the metal surface, which led to increasing the surface coverage and consequently protection efficiency even in the case of low inhibitor concentrations.

Okafor et al. looked into the extracts of onion (Allium sativum), Carica papaya extracts, Garcinia kola, andPhyllanthus amarus [19–22]. El-Etre, Abdallah M used Natural honey as corrosion inhibitor for metals and alloys. II C-steel in high saline water [23]. Jojoba oil has also been evaluated [24]. Artemisia oil has been investigated for it is anticorrosion properties [25]. Oguzie and coworkers evaluated Telfaria occidentalis,Occinum viridis, Azadirachta indica, and Sanseviera trifasciata extracts [26–29]. Benda-hou et al., studied using the extracts of rosemary in steel [27], and Sethuraman studied Datura [30]. Recently, studies on the use of some drugs as corrosion inhibitors have been reported by some researchers [31, 32]. Most of these drugs are heterocyclic compounds and were found to be environmentally friendly, hence, they have great potentials of competing with plant extracts. According to Eddy et al. drugs are environmentally friendly because they do not contain heavy metals or other toxic compound. In view of this adsorption and inhibitive efficiencies of ACPDQC (5-amino-1-cyclopropyl-7-[(3R, 5S) 3, 5-dimethylpiperazin-1-YL]-6,8-difluoro-4-oxo-uinoline-3-carboxylic acid), on mild steel corrosion have been studied and found to be effective.

Eddy et al. [33] studied inhibition of the corrosion of mild steel by ethanol extract of Musa species peel using hydrogen evolution and thermometric methods of monitoring corrosion. Inhibition efficiency of the extract was found to vary with concentration, temperature, period of immersion, pH, and electrode potentials. Adsorption of Musa species extract on mild steel surface was spontaneous and occurred according to Langmuir and Frumkin adsorption isotherms and also physical adsorption. Deepa Rani and Selvaraj [34] report the inhibition efficacy of Punica granatum extract on the corrosion of Brass in 1 N HCl evaluated by mass loss measurements at various time and temperature. Langmuir and Frumkin adsorption isotherms appear to be the mechanism of adsorption based on the values of activation energy, free energy of adsorption. Few researchers have summarized the effect of plant extracts on corrosion [35–38]. Efforts to find naturally organic substances or biodegradable organic materials to be used as corrosion inhibitors over the years have been intensified. Several reports are available on the various natural products used as green inhibitors as shown in Tables 1 and 2. Low-grade gram flour, natural honey, onion, potato, gelatin, plant roots, leaves, seeds, and flowers gums have been reported as good inhibitors. However, most of them have been tested on steel and nickel sheets. Although some studies have been performed on aluminum sheets, the corrosion effect is seen in very mild acidic or basic solutions (millimolar solutions).

Table 1: Green inhibitors used for corrosion inhibition of steel

Sl. no.	Metal	Inhibitor source	Active ingredient	References
(1)	Steel	Tamarind		[39]
(2)	Steel	Tea leaves		[40]
(3)	Steel	Pomegranate juice and peels		[41]
(4)	Steel	Emblica officinalis		[42]
(5)	Steel	Terminalia bellerica		[43]
(6)	Steel	Eucalyptus oil	Monomtrene 1,8-cineole	[44]
(7)		Rosemary		[45]
(8)	C-steel, Ni, Zn	Lawsonia extract (Henna)	Lawsone (2-hydroxy-1, 4-napthoquinone resin and tannin, coumarine, Gallic, acid, and sterols)	[46]
(9)	Mild steel	Gum exudate	Hexuronic acid, neutral sugar residues, volatile monoterpenes, canaric and related triterpene acids, reducing and nonreducing sugars	[47]
(10)	Mild steel	Musa sapientum peels (Banana peels)		[48]
(11)	Carbon steel	Natural amino acids—alanine, glycine, and leucine		[48]
(12)	Steel	Natural amino acids		[15]
(13)	Mild steel	Garcinia kola seed	Primary and secondary amines Unsaturated fatty acids and biflavnone	[49]
(14)	Steel	Auforpio turkiale	Protein hydrolysis	[50]
(15)	Steel	Azydracta indica	Protein hydrolysis	[51]
(16)	Steel	Aloe leaves		[52]
(17)	Steel	Mango/orange peels		[53]
(18)	Steel	Hibiscus sabdariffa (Calyx extract) in 1 M H_2SO_4 and 2 M HCl solutions, Stock 10–50%	Molecular protonated organic species in the extract. Ascorbic acid, amino acids, flavonoids, Pigments and carotene	[54]

Table 2: Green Inhibitors used for corrosion inhibition of aluminum, aluminum alloys, and other metals and alloys

Sl. no.	Metal	Inhibitor source	Active ingredient	References
(1)	Al	CeCl$_3$ and mercaptoben-zothiazole (MBT)		[55]
(2)	Al, steel	Aqueous extract of tobacco plant and its parts	Nicotine	[56]
(3)	Al	Vanillin		[57]
(4)	Al-Mg alloy	Aqueous extract of Rosmarinus officinalis—Neutral phenol subfraction of the aqueous extract	Catechin	[58]
(5)	Al	Sulphates/molybdates and dichromates as passivators		[59]
(6)	Al	Amino and polyamino acids—aspartic acid		[6]
(7)	Al	Pyridine and its selected derivatives (symmetric collidine and 2,5-dibrompyridine)		[60]
(8)	Al	Citric acid		[61]
(9)	Fe, Al	Benzoic acid		[62]
(10)	Al	Rutin and quercetin		[63]
(11)	Al			US Patent 5951747
(12)	Al	Polybutadieonic acid		[64]
(13)	Al and Zn	Saccharides—mannose and fructose		[65]
(14)	Al, Al-6061 and Al-Cu	Neutral solutions using sulphates, molybdates, and dichromates		[66]
(15)	Al	Vernonia amygdalina (Bitter leaf)		[67]
(16)	Al	Prosopis—cineraria (khejari)		[60]
(17)	Al	Tannin beetroot		[68]
(18)	Al	Saponin		[69]
(19)	Al	Acacia concianna		[70]
(20)	Al and Zn	Saccharides		[71]

(21)	Al	Opuntia (modified stems cladodes)	Polysaccharide (mucilage and pectin)	[72]
(22)	Al-Mg alloy	Rosmarinus officinalis		[8]
(23)	Zn	Metal chelates of citric acid		[61]
(24)	Zn	Onion juice	S-containing acids (glutamyl peptides) S-(1-propenyl) L-cysteine sulfoxide, and S-2-carboxy-propyl glutathione	[63]
(25)	Sn	Natural honey (acacia chestnut)		[64]
(26)	Sn	Black radish	120	[8]

Mechanism of Action of Green Inhibitors

Many theories to substantiate the mode of action of these green inhibitors have been put forth by several workers. Mann has suggested that organic substances, which form onium ions in acidic solutions, are adsorbed on the cathodic sites of the metal surface and interfere with the cathodic reaction.

Various mechanisms of action have been postulated for the corrosion inhibition property of the natural products.

Argemone Mexicana

It is a contaminant of mustard seeds contain an alkaloid berberine which has a long-chain of aromatic rings, an N atom in the ring, and, at several places H atoms attached to C are replaced by groups, – CH, –OCH$_3$, and –O. The free electrons on the O and N atoms form bonds with the electrons on the metal surface. Berberine in water ionizes to release a proton, thus the now negatively charged O atom helps to free an electron on the N atom and forms a stronger bond with the metallic electrons. These properties confer good inhibition properties to Argemone mexicana (Scheme 2).

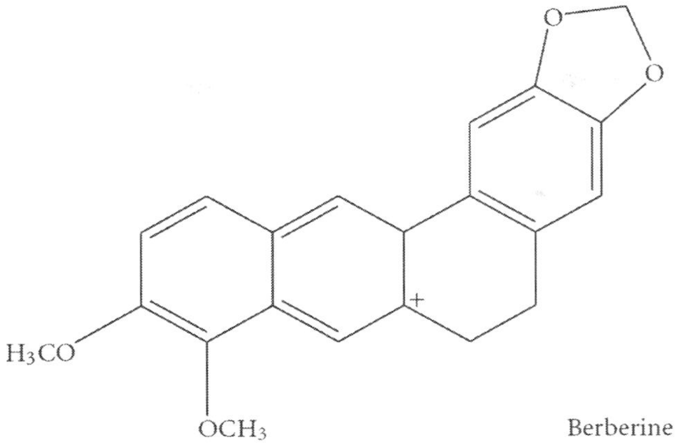

Berberine

Scheme 2: Berberine.

Garlic

It contains allyl propyl disulphide. Probably, this S-containing unsaturated compounds affects the potential cathodic process of steel.

Carrot

It contains pyrrolidine in aqueous media, pyrrolidine ionizes, and the N atom acquires a negative charge, and the free electrons on N possess still higher charge, resulting in stronger bond formation at N Carrot does not ionize in acidic media and thus does not protect in acids (Scheme 3).

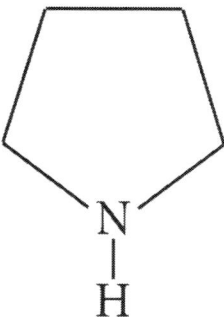

Pyrrolidine

Scheme 3: Pyrrolidine.

Castor Seed

They contain the alkaloid ricinine. The N atom is in the ring attachment of the $-OCH_3$ (Scheme 4).

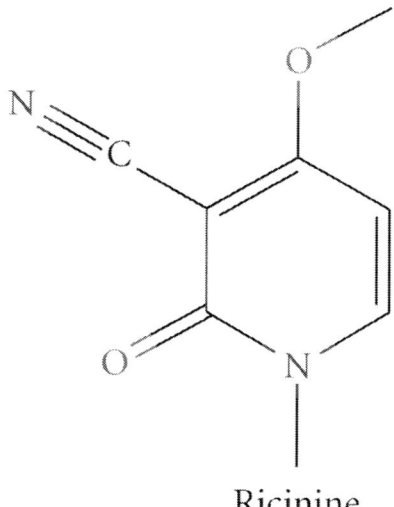

Ricinine

Scheme 4: Ricinine.

Black Pepper

Quraishi et al. [73] studied corrosion inhibition of mild steel in hydrochloric solution by black pepper extract (Piper nigrum family: Piperaceae) by mass loss measurements, potentiodynamic polarisation, and electrochemical impedance spectroscopy (EIS). Black pepper extract gave maximum inhibition efficiency (98%) at 120 ppm at 35°C for mild steel in hydrochloric acid medium. Electrochemical evaluation revealed it to be a mixed-type inhibitor and that charge transfer controls the corrosion process. The corrosion inhibition property was attributed to an alkaloid "Piperine".

Fennel Seeds

Essential oil from fennel (Foeniculum vulgare) (FM) was tested as corrosion inhibitor of carbon steel in 1 M HCl using electrochemical impedance spectroscopy (EIS), Tafel polarisation methods, and weight loss measurements [74]. The increase of the charge-transfer resistance () with the oil concentration supports the molecules of oil adsorption on the metallic surface. The polarization plots reveal that the addition of natural oil shifts the cathodic and anodic branches towards lower currents, indicative of a mixed-type inhibitor. The analysis of FM oil, obtained by hydrodistillation, using Gas Chromatography (GC) and Gas Chromatography/Mass Spectrometry (GC/MS) showed that the major components were limonene (20.8%) and pinene (17.8%). Interestingly, the composition of FM oil was variable according to the area of harvest and the stage of development. The analysis allowed the identification of 21 components which accounted for 96.6% of the total weight. The main constituents were limonene (20.8%) and pinene (17.8%) followed by myrcene (15%) and fenchone (12.5%). The adsorption of these molecules could take place via interaction with the vacant d-orbitals of iron atoms (chemisorption). It is logical to assume that such adsorption is mainly responsible for the good protective properties by a synergistic effect of various molecules [74–76].

Garcinia Mangostana

Vinod Kumar et al. [77] studied the corrosion inhibition of acid extract of the pericarp of the fruit of G. mangostana on mild steel in hydrochloric acid medium. G. mangostana, colloquially known as "the mangosteen", is a tropical evergreen tree. Mangosteen fruit, (Figure 1) on ripening the fruit, turns from green to purple in colour.

(a)

(b)

Figure 1: (a) Mangostana fruit. (b) Pericarp.

The extract of the pericarp of G. mangostana contains oxygenated prenylated xanthones, 8-hydroxycudraxanthone G and mangostingone [7-methoxy-2-(3- methyl-2-butenyl)-8-(3-methyl-2-oxo-3-butenyl)-1,3,6-trihydroxyxanthone, along with other xanthones such as cudraxanthone G, 8-deoxygartanin, garcimangosone B, garcinone D, garcinone E, gartanin, 1-isomangostin, α´-mangostin, γ-mangostin, mangostinone, smeathxanthone A, and tovophyllin A [77, 78]. Electrochemical parameters such as Ecorr, β_a, and c indicate the mixed mode of inhibition, but predominantly cathodic. IR analysis and impedance studies indicate that the adsorption on the metal surface is due to the heteroatoms present in the organic constituents of the extract of G. mangostana.

Ipomea Involcrata

Obot et al. [79] studied the corrosion inhibition efficiency of Ipomoea involcrata (IP) (family: Convolulaceae) leaf extract on aluminium. It is a common ornamental vine with heart-shaped and bright white pink or purple flowers that has a long history of use in central to southern Mexico. The plant has been shown to contain mainly d-lysergic acid amide (LSA) (Figure 2) and small amounts of other alkaloids, namely, chanoclavine, elymoclavine, and ergometrine, and d-isolysergic acid amide [79]. D-lysergic acid amide (LSA) (Figure 2) contains N and O in their structure including ϖ-electrons which are required for corrosion inhibiting effects. Probably, chanoclavine, elymoclavine, ergometrine, d-isolysergic acid amide, and other ingredients of the plant extracts synergistically increase the strength of the layer formed by the d-lysergic acid amide (LSA). Thus, the formation of a strong physisorbed layer between the metal surface and the phytoconstituents of the plant extract could be the cause of the inhibitive effect. The above authors have also reported that Chromolaena odorata as an excellent inhibitor for aluminium corrosion [80]. The environmentally friendly inhibitor could find possible applications in metal surface anodizing and surface coating in industries.

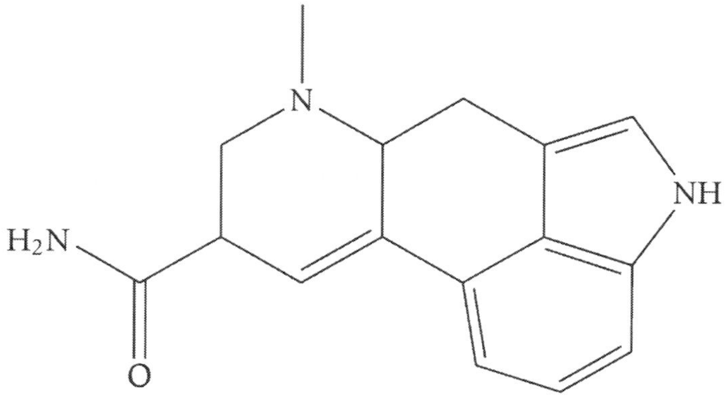

7-Methyl-4, 6.6a, 7, 8, 9-hexahydro-
indolo[4, 3-fg]quinoline-9-carboxamide

Figure 2: Structure of lysergic acid.

Soya Bean

It is rich in proteins, which are often good inhibitors in acidic media.

Most natural extracts constitute of oxygen- and nitrogen-containing compounds. Most of the oxygen-containing constituents of the extracts is a hydroxy aromatic compound, for example, tannins, pectins, flavonoids, steroids, and glycosides. Tannins are believed to form a passivating layer of tannates on the metallic surface. Similarly, it is postulated that a number of OH groups around the molecule lure them to form strong links with hydrogen and form complexes with metals. The complexes thus formed cause blockage of micro anodes and/or microanodes, which are generated on the metal surfaces when in contact with electrolytes, and, hence, retard subsequent dissolution of the metal.

Terminalia Catappa

The inhibitive and adsorption properties of ethanol extract of Terminalia catappa for the corrosion of mild steel in H_2SO_4 were investigated using weight loss, hydrogen evolution, and infrared methods of monitoring corrosion. The inhibition potential of ethanol extract of T. catappa is attributed to the presence of saponin, tannin, phlobatin, anthraquinone, cardiac glycosides, flavanoid, terpene, and alkaloid in the extract. The adsorption of the inhibitor on mild steel surface is exothermic, spontaneous, and best described by Langmuir adsorption model [81] similar results were reported for Gnetum Africana [82].

Caffeic Acid

De Souza and Spinelli [83] studied the inhibitory action of caffeic acid as a green corrosion inhibitor for mild steel. The inhibitor effect of the naturally occurring biological molecule caffeic acid on the corrosion of mild steel in $0.1 M$ H_2SO_4 was investigated by weight loss, potentiodynamic polarization, electrochemical impedance, and Raman spectroscopy. The different techniques confirmed the adsorption of caffeic acid onto the mild steel surface and consequently the inhibition of the corrosion process. Caffeic acid acts by decreasing the available cathodic reaction area and modifying the activation energy of the anodic reaction.

Gossypium Hirsutum

The corrosion inhibition properties of Gossypium hirsutum L leave extracts (GLE) and seed extracts (GSE) in $2 M$ sodium hydroxide (NaOH) solutions were studied using chemical technique. Gossypium extracts inhibited the corrosion of aluminium in NaOH solution. The inhibition efficiency increased with increasing concentration of the extracts. The leave extract (GLE) was found to be more effective than the seed extract (GSE). The GLE gave 97% inhibition efficiency while the GSE gave 94% at the highest concentration [83].

It is found that ethanol extract of M. sapientum peels (banana) can be used as an inhibitor for mild steel corrosion. The inhibitor acts by being adsorbed on mild steel surface according to classical adsorption models of Langmuir and Frumkin adsorption isotherms. Adsorption characteristics of the inhibitor follow physical adsorption mechanism. It is found that temperature, pH, period of immersion, electrode potential, and concentration of the inhibitor basically control the inhibitive action of M. sapientum peels.

Carmine and Fast Green Dyes

The use of dyes such as azo compounds methyl yellow, methyl red, and methyl orange [84] as inhibitors for mild steel has been reported [85–87]. The inhibition action of carmine and fast green dyes on corrosion of mild steel in 0.5 M HCl was investigated using mass loss, polarization, and electrochemical impedance (EIS) methods. Fast green showed inhibition efficiency of 98% and carmine 92%. The inhibitors act as mixed type with predominant cathodic effect.

Corrosion inhibition of mild steel in acidic solution by the dye molecules can be explained on the basis of adsorption on the metal surface, due to the donor-acceptor interaction between ϖ electrons of donor atoms N, O and aromatic rings of inhibitors, and the vacant d-orbitals of iron surface atoms [88, 89]. The fast green molecules possess electroactive nitrogen, oxygen atoms, and aromatic rings, favouring the adsorption while the carmine molecules possess electroactive oxygen atoms and electron rich paraquinanoid aromatic rings. In addition, the large and flat structure of the molecules occupies a large area of the substrate and thereby forming a protective coating. The inhibitors were adsorbed on the mild steel surface according to the Temkin adsorption isotherm (Figure 3).

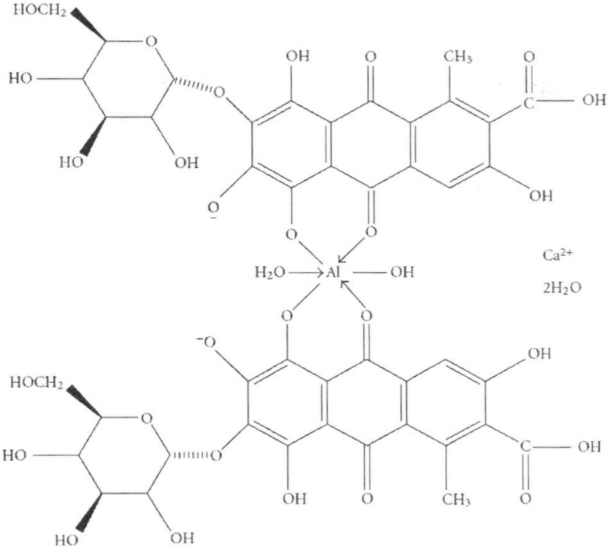

(a)

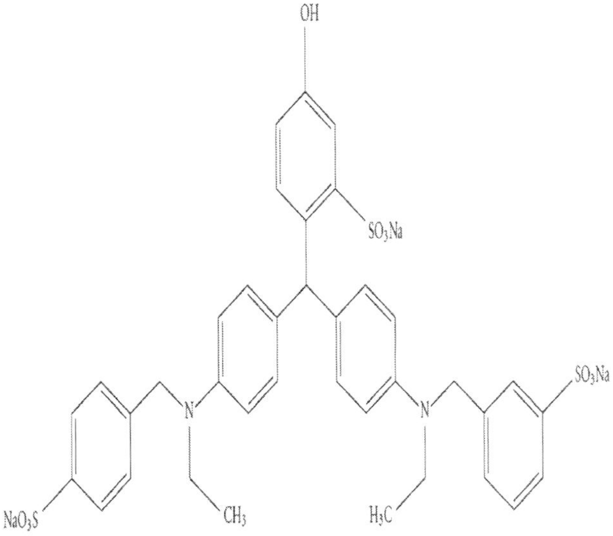

(b)

Figure 3: Structure of (a) carmine and (b) fast green.

Torres et al. [90] studied the effects of aqueous extracts of spent coffee grounds on the corrosion of carbon steel in a 1 mol L^{-1} HCl. Two methods of extraction were studied: decoction and infusion. The inhibition efficiency of C-steel in 1 mol L^{-1} HCl increased as the extract concentration and temperature increased. The coffee extracts acted as a mixed-type inhibitor with predominant cathodic effectiveness. In this study, the adsorption process of components of spent coffee grounds extracts obeyed the Langmuir adsorption isotherm. The chlorogenic acids isolated do not appear to be the active ingredient.

Biocorrosion and Prevention by Green Inhibitors

Biocorrosion relates to the presence of microorganisms that adhere to different industrial surfaces and damage the metal. Bacterial cells encase themselves in a hydrated matrix of polysaccharides and protein and form a slimy layer known as biofilm. The biofilm is a gel consisting of approximately 95% water, microbial metabolic products like enzymes, extracellular polymeric substances, organic and inorganic acids, and also volatile compounds such as ammonia or hydrogen sulphide and inorganic detritus [90–92]. Extracellular polymeric substances play a crucial role in biofilm development. Inhibition of biofilm formation is the simplest way of biocorrosion prevention. Use of naturally produced compounds such as plant extracts could be used as effective biocides [34].

SOL-GEL COATINGS

In recent years, the sol-gel coatings doped with inhibitors developed to replace chromate conversion coatings show real promise [93]. Results show that the corrosion resistance of the sol-gel coatings containing CeCl$_3$ proves to be better than that of the pure and MBT-added sol-gel coatings by the electrochemical methods. However, unlike chromium, silane-based sol-gel coatings mainly act as physical barrier rather than form chemical bond with substrate. Inhibitors are necessary to release in the coating film to slow the corrosion process through self-healing effect [57, 89, 94–96]. Among the inhibitors, rare-earth elements are generally considered to be effective and nontoxic in sol-gel coatings. Additionally, some organic inhibitors, especially heterocyclic compounds, are effective as slowly released inhibitors in sol-gel coating [97, 98]. Andreeva et al. suggested self-healing anticorrosion coatings based on pH [99, 100]. The approach to prevention of corrosion propagation on metal surfaces achieving the self-healing effect is based on suppression of accompanying physicochemical reactions. The corrosion processes are followed by changes of the pH value in the corrosive area and metal degradation. Self-healing or self-curing of the areas damaged by corrosion can be performed by three mechanisms: pH neutralization, passivation of the damaged metal surface by inhibitors entrapped between polyelectrolyte layers, and repair of the coating. The corrosion inhibitor incorporated as a component of the layer-by-layer film into the protective coating is responsible for the most effective mechanism of corrosion suppression. Quinolines are environmentally friendly corrosion inhibitors that are attracting more and more attention as alternatives to the harmful chromates.

Recent awareness of the corrosion inhibiting abilities of tannins, alkaloids, organic and amino acids, as well as organic dyes has resulted in sustained interest on the corrosion inhibiting properties of natural products of plant origin. Such investigation is of much importance because in addition to being environmentally friendly and ecologically acceptable, plant products are inexpensive, readily available, and renewable sources of materials. Although a number of insightful papers have been devoted to corrosion inhibition by plant extracts, reports on the detailed mechanisms of the adsorption process are still scarce. The drawback of most reports on plant extracts as corrosion inhibitors is that the active ingredient has not been identified.

In recent years, sol-gel coatings doped with green inhibitors show real promise for corrosion protection of a variety of metals and alloys.

COMPUTATIONAL MODELING FOR CORROSION

Simulation is a prognostic computational tool for complex scientific and engineering problems. The simplest simulation methods have been used for decades, but, with the increase in computational memory and speed simulation, have become the prevalent tool for analysis [101–103]. Simulation turns probability models into statistics problems where the results can be analyzed using standard statistical methods. The challenge of a simulation is to implement a procedure that efficiently captures the desired model characteristics. Often the goal of probability computations is the evaluation of high reliability. In fact, computation of high reliabilities itself is an ongoing research concern. Hence, there is no one way in which to do the computation. Monte Carlo simulation is the traditional and powerful method if computational complexity and time are not limiting. The Box-Muller method is also well known. A variety of techniques have been developed to reduce the number of simulations without compromising accuracy.

The study of corrosion involves the study of the chemical, physical, metallurgical, and mechanical properties of materials as it is a synergistic phenomenon in which the environment is as equally important as the materials involved. Computer modeling techniques can handle the study of complex systems such as corrosion and thus are appropriate and powerful tools to study the mechanism of action of corrosion and its inhibitors.

In the recent past, computer modeling techniques have been successfully applied to corrosion problems as summarized in review articles by Zamani et al. [104] and Munn [105]. The application of computer modeling techniques to corrosion systems requires an understanding of the physical phenomenon of corrosion and the mathematics which govern the corrosion process. In addition, knowledge of the numerical procedures which are the basis of computer modeling techniques is essential for accurate computational analyses. In addition, validation of the computer analysis results with experimental data is mandatory. Without a reasonably accurate description of the damage process at a scale that is pertinent to the desired application, probabilistic computations have minimal value for prognosis and life-cycle assessment.

For corrosion modeling, the materials characterization depends on the orientation of the material. Figure 4 is a composite of three optical micrographs of the perpendicular faces of a typical specimen of 7075-T6 aluminum alloy, where LT, LS, and TS are the rolling, long-transverse, and short-transverse planes, respectively. Visually, there is a difference in the three surfaces, and the variability in the location, size, and density of the particles is apparent. Thus, for eg when modeling for aircraft wings, the LS surface is the most significant surface to characterize because it is the surface in fastener holes subjected to high-stress loading.

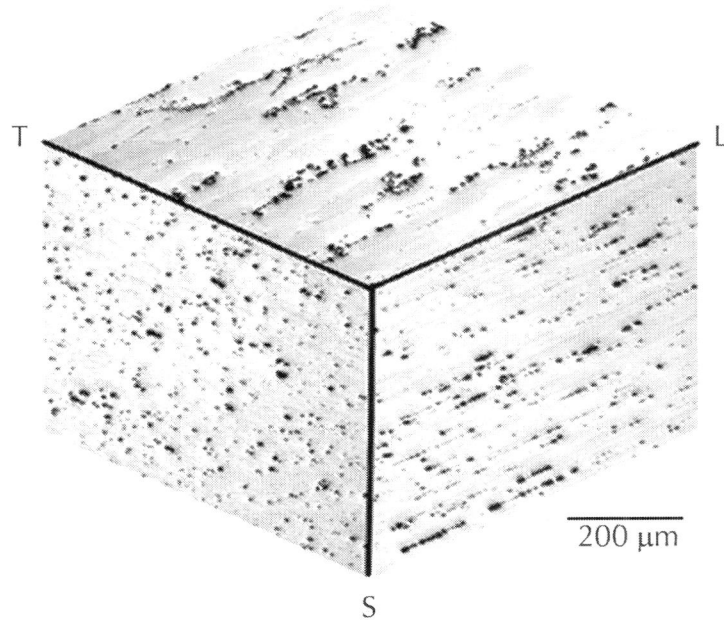

Figure 4: Three optical micrographs of the perpendicular faces of a typical specimen of 7075-T6 aluminum alloy.

Some Examples of Computational Modeling in Corrosion Inhibition

Tryptophan

According to the description of frontier orbital theory, HOMO is often associated with the electron donating ability of an inhibitor molecule. High EHOMO values indicate that the molecule has a tendency to donate electrons to the metal with unoccupied molecule orbitals. ELUMO indicates the ability of the molecules to accept electrons. The lower value of ELUMO is the easier acceptance of electrons from metal surface. The gap between the LUMO and HOMO energy levels of the inhibitor molecules is another important index, and the low absolute values of the energy band gap (DE = ELUMO − EHOMO) means good inhibition efficiency. Studies indicated that L-tryptophan has high value of EHOMO and low value of ELUMO with low-energy band gap. Adsorption energy calculated for the adsorption of L-tryptophan on Fe surface in the presence of water molecules equals −29.5 kJ mol^{-1}, which implies that the interaction between L-tryptophan molecule and Fe surface is strong [105, 106]. Molecule dynamics simulation results showed that L-tryptophan molecules assumed a nearly flat orientation with respect to the Fe (1 1 0) surface. The calculated adsorption energy between a L-tryptophan molecule and Fe surface is −29.5 kJ mol^{-1}.

The optimized molecule structure, the highest occupied molecule orbitals, the lowest unoccupied molecule orbital, and the charge distribution of L-tryptophan molecule using DFT functional (B3LYP/6-311*G) are shown in Figure 5. The figure shows that in L-tryptophan molecule, C5, C12, C13, C14, C15, N7, N10, O2, and O4 carry more negative charges, while C8 and C6 carry more positive charges.

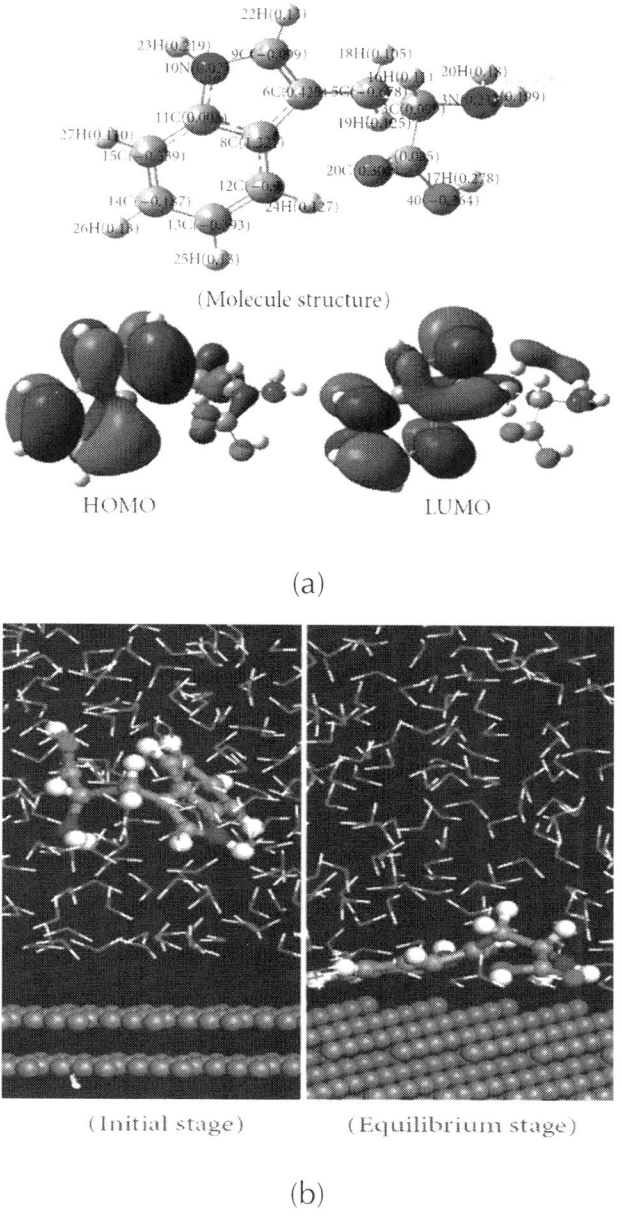

(a)

(Initial stage)   (Equilibrium stage)

(b)

Figure 5: (a) Optimised molecule structure and charge density distribution of L-tryptophan. (b) L-tryptophan adsorbed on Fe surface in water solution.

This means that C5, C12, C13, C14, C15, N7, N10, O2, and O4 are the negative charge centers, which can offer electrons to the Fe atoms to form coordinate bond, and C8 and C6 are the positive charge centers, which can accept electrons from orbital of Fe atoms to form feedback bond. The optimized structure is in accordance with the fact that excellent corrosion inhibitors cannot only offer electrons to unoccupied orbital of the metal, but also accept free electrons from the metal. Therefore, it can be inferred that indole ring, nitrogen, and oxygen atoms are the possible active adsorption sites.

Presuel-Moreno et al. [107] modeled the chemical throwing power of an Al-Co-Ce metallic coating under thin electrolyte films representative of atmospheric conditions. An Al-Co-Ce alloy coating was developed for an AA2024-T3 substrate that can serve as barrier, sacrificial anode, and reservoir to supply soluble inhibitor ions to protect any defect sites or simulated scratches exposing the substrate. The model calculates the time necessary to accumulate Ce +3 and Co +2 inhibitors over the scratch when released from the Al-Co-Ce coating under different conditions such as the pH-dependent passive dissolution rate of an Al-Co-Ce alloy to define the inhibitor release flux. Transport by both electromigration and diffusion was considered. The effects of scratch size, initial pH, chloride concentration, and electrochemical kinetics of the material involved were studied. Studies indicated that sufficient accumulation of the released inhibitor (i.e., the Ce +3 concentration surpassed the critical inhibitor concentration over AA2024-T3 scratches) was achieved within a few hours (e.g., ~4 h for scratches of S = 1500 μm) when the initial solution pH was 6 and the coating was adjacent to the AA2024-T3.

Pradip and Rai [108] modeled design of phosphonic-acid-based corrosion inhibitors using a force field approach.

Piperidine and Derivatives

Khaled and Amin [109] studied the adsorption and corrosion inhibition behaviour of four selected piperidine derivatives, namely, piperidine (pip), 2-methylpiperidine (2mp), 3-methylpiperidine (3mp), and 4-methylpiperidine (4mp) at nickel in 1.0 M HNO_3 solution computationally by the molecular dynamics simulation and quantum chemical calculations and electrochemically by Tafel and impedance methods. The molecular dynamics (MD) simulations were performed using the commercial software MS Modeling from Accelrys using the amorphous cell module to create solvent piperidines cells on the nickel substrate. The behaviour of the inhibitors on the surface was studied using molecular dynamics simulations, and the condensed phase optimized molecular potentials for atomistic simulation studies (COMPASS) force field. COMPASS is an ab initio powerful force field which supports atomistic simulations of condensed phase materials [102]. Molecular simulation studies were applied to optimize the adsorption structures of piperidine derivatives. The nickel/inhibitor/solvent interfaces were simulated, and the charges on the inhibitor molecules as well as their structural parameters were calculated in the presence of solvent effects. Quantum chemical calculations based on the ab initio method were performed to determine the relationship between the molecular structure of piperidines and their inhibition efficiency. Results obtained from Tafel and impedance methods are in good agreement and confirm theoretical studies.

Khaled and Amin [110] also conducted studies on the molecular dynamics simulation on the corrosion inhibition of aluminum in molar hydrochloric acid using some imidazole derivatives. They also adapted Monte Carlo simulations technique incorporating molecular mechanics and dynamics to simulate the adsorption of methionine derivatives, namely, L-methionine, L-methionine sulphoxide, and L-methionine sulphone on iron (110) surface in 0.5 M sulphuric acid. Results show that methionine derivatives have a very good inhibitive effect for corrosion of mild steel in 0.5 M sulphuric acid solution.

Aniline and Its Derivatives

The inhibiting action of aniline and its derivatives on the corrosion of copper in hydrochloric acid has been investigated by Henriquez et al. [39], with emphasis on the role of substituents. With this purpose five different anilines were selected: aniline, p-chloroaniline, p-nitro aniline, p-methoxy, and p-methylaniline. A theoretical study using molecular mechanic and ab initio Hartree Fock methods, to model the adsorption of aniline on copper (100) showed results in good agreement with the experimental data. Aniline adsorbs parallel to the copper surface, showing no preference for a specific adsorption site. On the other hand, from ab initio Hartree Fock calculations, adsorption energy between 2 kcal/mol and 5 kcal/mol is obtained, which is close to the experimental value, confirming that the adsorption of aniline on the metal substrate is rather weak. In view of these results, the orientation of the aniline molecule with respect to the copper surface is considered to be the dominant effect. Mechanic molecular calculations were carried out using the Insight II, a comprehensive graphic molecular modeling program, to obtain configurations of minimum energy.

ACKNOWLEDGMENTS

The encouragement and cooperation received from Dr. Upadhya, Director, NAL, Bangalore, Dr. Ranjan Moodithaya, Head, KTMD, and Dr. K.S. Rajam, Head, SED are gratefully acknowledged. Patents used in the paper are: (1) US Patent 5951747—Non-chromate corrosion inhibitors for aluminum alloys; (2) United States Patent 5286357—Corrosion sensors; (3) WO/2002/008345—CORROSION INHIBITORS; and (4) British patent, 2327,1895.

REFERENCES

1.   M. Bouklah, B. Hammouti, T. Benhadda, and M. Benkadour, "Thiophene derivatives as effective inhibitors for the corrosion of steel in 0.5 M H_2SO_4," Journal of Applied Electrochemistry, vol. 35, no. 11, pp. 1095–1101, 2005. View at Google Scholar

2.   A. S. Fouda, A. A. Al-Sarawy, and E. E. El-Katori, "Pyrazolone

derivatives as corrosion inhibitors for C-steel HCl solution," Desalination, vol. 201, pp. 1–13, 2006.

3. A. Fiala, A. Chibani, A. Darchen, A. Boulkamh, and K. Djebbar, "Investigations of the inhibition of copper corrosion in nitric acid solutions by ketene dithioacetal derivatives," Applied Surface Science, vol. 253, no. 24, pp. 9347–9356, 2007.

4. U. R. Evans, The Corrosion and Oxidation of Metals, Hodder Arnold, 1976.

5. O. K. Abiola, N. C. Oforka, E. E. Ebenso, and N. M. Nwinuka, "Eco-friendly corrosion inhibitors: The inhibitive action of Delonix Regia extract for the corrosion of aluminium in acidic media," Anti-Corrosion Methods and Materials, vol. 54, no. 4, pp. 219–224, 2007.

6. M. Kliskic, J. Radoservic, S. Gudic, and V. Katalinic, "Aqueous extract of Rosmarinus officinalis L. as inhibitor of Al-Mg alloy corrosion in chloride solution," Journal of Applied Electrochemistry, vol. 30, no. 7, pp. 823–830, 2000.

7. A. Y. El-Etre, "Natural honey as corrosion inhibitor for metals and alloys. I. Copper in neutral aqueous solution," Corrosion Science, vol. 40, no. 11, pp. 1845–1850, 1998.

8. A. Y. El-Etre, "Inhibition of aluminum corrosion using Opuntia extract," Corrosion Science, vol. 45, no. 11, pp. 2485–2495, 2003. View at Publisher · · View at Scopus

9. A. Y. El-Etre, "Khillah extract as inhibitor for acid corrosion of SX 316 steel," Applied Surface Science, vol. 252, no. 24, pp. 8521–8525, 2006.

10. E. E. Ebenso, U. J. Ibok, U. J. Ekpe et al., "Corrosion inhibition studies of some plant extracts on aluminium in acidic medium," Transactions of the SAEST, vol. 39, no. 4, pp. 117–123, 2004

11. E. E. Ebenso and U. J. Ekpe, "Kinetic study of corrosion and corrosion inhibition of mild steel in H_2SO_4 using Parica papaya leaves extract," West African Journal of Biological and Applied Chemistry, vol. 41, pp. 21–27, 1996.

12. U. J. Ekpe, E. E. Ebenso, and U. J. Ibok, "Inhibitory action of Azadirachta indica leaves extract on the corrosion of mild steel in H_2SO_4," West African Journal of Biological and Applied Chemistry, vol. 37, pp. 13–30, 1994.

13. F. Zucchi and I. H. Omar, "Plant extracts as corrosion inhibitors of mild steel in HCl solutions," Surface Technology, vol. 24, no. 4, pp. 391–399, 1985

14. S. A. Umoren, O. Ogbobe, I. O. Igwe, and E. E. Ebenso, "Inhibition of mild steel corrosion in acidic medium using synthetic and naturally occurring polymers and synergistic halide additives," Corrosion Science, vol. 50, no. 7, pp. 1998–2006, 2008.

15. Y. J. Yee, Green inhibitors for corrosion control: a Study on the inhibitive effects of extracts of honey and rosmarinus officinalis L. (Rosemary), M.S. thesis, University of Manchester, Institute of Science and Technology, 2004.

16. J. C. Chalchat, R. P. Garry, A. Michet, B. Benjilali, and J. L. Chabart, "Essential oils of Rosemary (Rosmarinus officinalis L.). The chemical composition of oils of various origins (Morocco, Spain, France)," Journal of Essential Oil Research, vol. 5, no. 6, pp. 613–618, 1993.

17. E. El Ouariachi, J. Paolini, M. Bouklah et al., "Adsorption properties of Rosmarinus of ficinalis oil as green corrosion inhibitors on C38 steel in 0.5 M H_2SO_4," Acta Metallurgica Sinica, vol. 23, no. 1, pp. 13–20, 2010.

18. A. O. Odiongenyi, S. A. Odoemelam, and N. O. Eddy, "Corrosion inhibition and adsorption properties of ethanol extract of Vernonia Amygdalina for the corrosion of mild steel in H_2SO_4," Portugaliae Electrochimica Acta, vol. 27, no. 1, pp. 33–45, 2009.

19. S. A. Umoren and E. E. Ebenso, "Studies of the anti-corrosive effect of Raphia hookeri exudate gum-halide mixtures for aluminium corrosion in acidic medium," Pigment and Resin Technology, vol. 37, no. 3, pp. 173–182, 2008.

20. M. Abdallah, "Guar gum as corrosion inhibitor for carbon

steel in sulphuric acid solutions," Portugaliae Electrochimica Acta, vol. 22, pp. 161–175, 2004.

21. P. C. Okafor, U. J. Ekpe, E. E. Ebenso, E. M. Umoren, and K. E. Leizou, "Inhibition of mild steel corrosion in acidic medium by Allium sativum extracts," Bulletin of Electrochemistry, vol. 21, no. 8, pp. 347–352, 2005.

22. P. C. Okafor and E. E. Ebenso, "Inhibitive action of Carica papaya extracts on the corrosion of mild steel in acidic media and their adsorption characteristics," Pigment and Resin Technology, vol. 36, no. 3, pp. 134–140, 2007.

23. P. C. Okafor, V. I. Osabor, and E. E. Ebenso, "Eco-friendly corrosion inhibitors: Inhibitive action of ethanol extracts of Garcinia kola for the corrosion of mild steel in H_2SO_4 solutions," Pigment and Resin Technology, vol. 36, no. 5, pp. 299–305, 2007.

24. P. C. Okafor, M. E. Ikpi, I. E. Uwah, E. E. Ebenso, U. J. Ekpe, and S. A. Umoren, "Inhibitory action of Phyllanthus amarus extracts on the corrosion of mild steel in acidic media," Corrosion Science, vol. 50, no. 8, pp. 2310–2317, 2008.

25. A. Y. El-Etre and M. Abdallah, "Natural honey as corrosion inhibitor for metals and alloys. II. C-steel in high saline water," Corrosion Science, vol. 42, no. 4, pp. 731–738, 2000.

26. A. Chetouani, B. Hammouti, and M. Benkaddour, "Corrosion inhibition of iron in hydrochloric acid solution by jojoba oil," Pigment and Resin Technology, vol. 33, no. 1, pp. 26–31, 2004.

27. A. Bouyanzer and B. Hammouti, "A study of anti-corrosive effects of Artemisia oil on steel," Pigment and Resin Technology, vol. 33, no. 5, pp. 287–292, 2004.

28. E. E. Oguzie, "Inhibition of acid corrosion of mild steel by Telfaria occidentalis extract," Pigment and Resin Technology, vol. 34, no. 6, pp. 321–326, 2005.

29. E. E. Oguzie, "Studies on the inhibitive effect of Occimum viridis extract on the acid corrosion of mild steel," Materials Chemistry and Physics, vol. 99, pp. 441–446, 2006.

30. E. E. Oguzie, "Corrosion inhibition of aluminium in acidic and alkaline media by Sansevieria trifasciataextract," Corrosion Science, vol. 49, no. 3, pp. 1527–1539, 2007.

31. M. A. Bendahou, M. B. E. Benadellah, and B. B. Hammouti, "A study of rosemary oil as a green corrosion inhibitor for steel in 2 M H_3PO_4," Pigment and Resin Technology, vol. 35, no. 2, pp. 95–100, 2006.

32. M. G. Sethuraman and P. B. Raja, "Corrosion inhibition of mild steel by Datura metel in acidic medium," Pigment and Resin Technology, vol. 34, no. 6, pp. 327–331, 2005.

33. N. O. Eddy, S. A. Odoemelam, and A. O. Odiongenyi, "Ethanol extract of musa species peels as a green corrosion inhibitor for mild steel: Kinetics, adsorption and thermodynamic considerations," Electronic Journal of Environmental, Agricultural and Food Chemistry, vol. 8, no. 4, pp. 243–255, 2009.

34. P. Deepa Rani and S. Selvaraj, "Inhibitive and adsorption properties of punica granatum extract on brass in acid media," Journal of Phytology, vol. 2, no. 11, pp. 58–64, 2010.

35. S. Rajendran, V. Ganga Sri, J. Arockiaselvi, and A. J. Amalraj, "Corrosion inhibition by plant extracts—an overview," Bulletin of Electrochemistry, vol. 21, no. 8, pp. 367–377, 2005.

36. K. Srivastava and P. Srivastava, "Studies on plant materials as corrosion inhibitors," British Corrosion Journal, vol. 16, no. 4, pp. 221–223, 1981.

37. R. M. Saleh, A. A. Ismail, and A. A. El Hosary, "corrosion inhibition by naturally occurring substances. vii. the effect of aqueous extracts of some leaves and fruit peels on the corrosion of steel, Al, Zn and Cu in acids," British Corrosion Journal, vol. 17, no. 3, pp. 131–135, 1982.

38. P. B. Raja and M. G. Sethuraman, "Natural products as corrosion inhibitor for metals in corrosive media—a review," Materials Letters, vol. 62, no. 1, pp. 113–116, 2008.

39. J. H. Henriquez-Román, M. Sancy, M. A. Páez et al., "The

influence of aniline and its derivatives on the corrosion behaviour of copper in acid solution," Journal of Solid State Electrochemistry, vol. 9, no. 7, pp. 504–511, 2005.

40. A. A. El Hosary, R. M. Saleh, and A. M. Shams El Din, "Corrosion inhibition by naturally occurringsubstances-I. The effect of Hibiscus subdariffa (karkade) extract on the dissolution of Al and Zn," Corrosion Science, vol. 12, no. 12, pp. 897–904, 1972.

41. R. M. Saleh and A. M. Shams El Din, "Efficiency of organic acids and their anions in retarding the dissolution of aluminium," Corrosion Science, vol. 12, no. 9, pp. 689–697, 1972.

42. M. J. Sanghvi, S. K. Shuklan, A. N. Misra, M. R. Padh, and G. N. Mehta, "Inhibition of hydrochloric acid corrosion of mild steel by aid extracts of Embilica officianalis, Terminalia bellirica and Terminalia chebula," Bulletin of Electrochemistry, vol. 13, no. 8-9, pp. 358–361, 1997.

43. M. J. Shangvi, S. K. Shukla, A. N. Mishra, M. R. padh, and G. N. Mehta, "Corrosion inhibition of mild steel in hydrochloric acid by acid extracts of Sapindus Trifolianus, Acacia Concian and Trifla,"Transactions of the Metal Finishers Association of India, vol. 5, no. 3, pp. 143–147, 1996.

44. A. Chetouani and B. Hammouti, "Corrosion inhibition of iron in hydrochloric acid solutions by naturally henna," Bulletin of Electrochemistry, vol. 19, no. 1, pp. 23–25, 2003.

45. B. Muller, W. Klager, and G. Kubitzki, "Metal chelates of citric acid as corrosion inhibitors for zinc pigment," Corrosion Science, vol. 39, no. 8, pp. 1481–1485, 1997.

46. A. Bouyanzer, L. Majidi, and B. Hammouti, "Effect of eucalyptus oil on the corrosion of steel in 1M HCl," Bulletin of Electrochemistry, vol. 22, no. 7, pp. 321–324, 2006.

47. A. Y. El-Etre, "Natural onion juice as inhibitor for zinc corrosion," Bulletin of Electrochemistry, vol. 22, no. 2, pp. 75–80, 2006.

48. I. Radojcic, K. Berkovi , S. Kova , and J. Vorkapi -Fura , "Natural honey and black radish juice as tin corrosion inhibitors,"

Corrosion Science, vol. 50, no. 5, pp. 1498–1504, 2008.

49.   A. Y. El-Etre, M. Abdallah, and Z. E. El-Tantawy, "Corrosion inhibition of some metals using lawsonia extract," Corrosion Science, vol. 47, no. 2, pp. 385–395, 2005.

50.   S. A. Umoren, I. B. Obot, and E. E. Ebenso, "Corrosion inhibition of aluminium using exudate gum from Pachylobus edulis in the presence of halide ions in HCl," E-Journal of Chemistry, vol. 5, no. 2, pp. 355–364, 2008.

51.   N. O. Eddy and E. E. Ebenso, "Adsorption and inhibitive properties of ethanol extracts of Musa sapientum peels as a green corrosion inhibitor for mild steel in H_2SO_4," African Journal of Pure and Applied Chemistry, vol. 2, no. 6, pp. 046–054, 2008.

52.   S. Lyon, "A natural way to stop corrosion," Nature, vol. 427, no. 406, p. 407, 2004.

53.   E. E. Oguzie, K. L. Iyeh, and A. I. Onuchukwu, "Inhibition of mild steel corrosion in acidic media by aqueous extracts from Garcinia kola seed," Bulletin of Electrochemistry, vol. 22, no. 2, pp. 63–68, 2006.

54.   R. M. Saldo, A. A. Ismail, and A. A. El Hosary, "Corrosion Inhibition by naturally occurring substances," British Corrosion Journal, vol. 17, no. 3, pp. 131–135, 1990.

55.   E. E. Oguzie, "Corrosion inhibitive effect and adsorption bBehaviour of Hibiscus Sabdariffa extract on mild steel in acidic media," Portugaliae Electrochimica Acta, vol. 26, pp. 303–314, 2008.

56.   E. E. Oguzie, "Corrosion inhibitive effect and adsorption behaviour of Hibiscus sabdariffa extract on mild steel in acidic media," Portugaliae Electrochimica Acta, vol. 26, no. 3, pp. 303–314, 2008.

57.   H. W. Shi, F. C. Liu, E. H. Han, and M. C. Sun, "Investigation on a sol-gel coating containing inhibitors on 2024-T3 aluminum alloy," Chinese Journal of Aeronautics, vol. 19, pp. S106–S112, 2006.

58.   G. D. Davis, "The Use of Extracts of Tobacco Plants as

Corrosion   Inhibitors,"http://www.electrochem.Org/dl/ma/202/pdfs/0340.PDF.

59. A. Y. El-Etre, "Inhibition of acid corrosion of aluminum using vanillin," Corrosion Science, vol. 43, no. 6, pp. 1031–1039, 2001.

60. W. A. Badawy, F. M. Allhara, and A. S. ElAzab, "Electrochemical behaviour and corrosion inhibition of Al, Al-6061 and Al-Cu in neutral aqueous solutions," Corrosion Science, vol. 41, no. 4, pp. 709–727, 1999.

61. B. Müller, "Amino and polyamino acids as corrosion inhibitors for aluminium and zinc pigments,"Pigment and Resin Technology, vol. 31, no. 2, pp. 84–87, 2002.

62. M. Kliskic, J. Radosevi, and S. Gudic, "Pyridine and its derivatives as inhibitors of aluminium corrosion in chloride solution," Journal of Applied Electrochemistry, vol. 27, no. 6, pp. 947–952, 1997.

63. R. Solmaz, G. Karda , B. Yazici, and M. Erbil, "Citric acid as natural corrosion inhibitor for aluminium protection," Corrosion Engineering Science and Technology, vol. 43, no. 2, pp. 186–191, 2008.

64. Z. Sibel, "The effects of benzoic acid in chloride solutions on the corrosion of iron and aluminum,"Turkish Journal of Chemistry, vol. 26, no. 3, pp. 403–408, 2002.

65. K. Berkovic, S. Kovac, and J. Vorkapic-Furac, "Natural compounds as environmentally friendly corrosion inhibitors of aluminium," Acta Alimentaria, vol. 33, no. 3, pp. 237–247, 2004.

66. Y. Tao, X. Zhang, and Z. Gu, "The inhibition of corrosion of aluminum in acid environment byin situ electrocoagulation of polybutadienoic acid," Wuhan University Journal of Natural Sciences, vol. 3, no. 2, pp. 221–225, 1998.

67. B. Müller and M. Kurfeß, "Saccharide und deren Derivate als Korrosionsinhibitoren für Aluminiumpigmente in wäßrigen Medien," Materials and Corrosion, vol. 44, no. 9, pp. 373–378, 2004.

68. G. O. Avwiri and F. O. Igho, "Inhibitive action of Vernonia amygdalina on the corrosion of aluminium alloys in acidic media," Materials Letters, vol. 57, no. 22-23, pp. 3705–3711, 2003. ·

69. S. Manish Kumar, K. Sudesh, R. Ratnani, and S. P. Mathur, "Corrosion inhibition of Aluminium by extracts of Prosopis cineraria in acidic media," Bulletin of Electrochemistry, vol. 22, no. 2, pp. 69–74, and 2006.

70. A. A. Hossary, M. M. Gauish, and R. M. Saleh, "Corrosion inhibitor formulations from coal-tar distillation products for acid cleaning of steel in HCl," in Proceedings of the 2nd International Symposium on Industrial and Oriented Basic Electrochemistry, pp. 6–18, SAEST, CECRI, Madras, India, 1980.

71. R. A. Tupikov, Y. G. Dragunov, I. L. Kharina, and D. S. Zmienko, "Protection of carbon steels against atmospheric corrosion in a wet tropical climate using gas-plasma metallization with aluminum," Protection of Metals, vol. 44, no. 7, pp. 673–682, 2008. ·

72. P. Arora, T. Jain, and S. P. Mathur, Chemistry, vol. 1, p. 766, 2005.

73. M. A. Quraishi, D. K. Yadav, and I. Ahamad, "Green approach to corrosion inhibition by black pepper extract in hydrochloric acid solution," Open Corrosion Journal, vol. 2, pp. 56–60, 2009.

74. N. Lahhit, A. Bouyanzer, J.-M. Desjobert, et al., "Fennel (Foeniculum vulgare) essential oil as green corrosion Inhibitor of carbon steel in hydrochloric acid solution," Portugaliae Electrochimica Acta, vol. 29, no. 2, pp. 127–138, 2011.

75. E. E. Oguzie, "Corrosion inhibition of mild steel in hydrochloric acid solution by methylene blue dye," Materials Letters, vol. 59, no. 8-9, pp. 1076–1079, 2005. ·

76. A. Zarrouk, I. Warad, B. Hammouti, A. Dafali, S. S. Al-Deyab, and N. Benchat, "The effect of temperature on the corrosion of Cu/HNO_3 in the Presence of organic inhibitor: Part-2,"

International Journal of Electrochemical Science, vol. 5, no. 10, pp. 1516–1526, 2010.

77. K. P. Vinod Kumar, M. S. Narayanan Pillai, and G. Rexin Thusnavis, "Pericarp of the fruit of garcinia mangostana as corrosion inhibitor for mild steel in hydrochloric acid medium," Portugaliae Electrochimica Acta, vol. 28, no. 6, pp. 373–383, 2010.

78. H. A. Jung, B. N. Su, W. J. Keller, R. G. Mehta, and A. D. Kinghorn, "Antioxidant xanthones from the pericarp of Garcinia mangostana (Mangosteen)," Journal of Agricultural and Food Chemistry, vol. 54, no. 6, pp. 2077–2082, 2006. · · View at PubMed

79. I. B. Obot, N. O. Obi-Egbedi, S. A. Umoren, and E. E. Ebenso, "Synergistic and antagonistic effects of anions and ipomoea invulcrata as green corrosion inhibitor for aluminium dissolution in acidic medium," International Journal of Electrochemical Science, vol. 5, no. 7, pp. 994–1007, 2010.

80. I. B. Obot and N. O. Obi-Egbedi, "Ipomoea involcrata as an ecofriendly inhibitor for aluminium in alkaline medium," Portugaliae Electrochimica Acta, vol. 27, no. 4, pp. 517–524, 2009.

81. I. B. Obot and N. O. Obi-Egbedi , "An interesting and efficient green corrosion inhibitor for aluminium from extracts of Chlomolaena odorata L. in acidic solution," Journal of Applied Electrochemistry, vol. 40, no. 11, pp. 1977–1983, 2010.

82. N. O. Eddy, P. A. Ekwumemgbo, and P. A. P. Mamza, "Ethanol extract of Terminalia catappa as a green inhibitor for the corrosion of mild steel in H_2SO_4," Green Chemistry Letters and Reviews, vol. 2, no. 4, pp. 223–231, 2009.

83. F. A. de Souza and A. Spinelli, "Caffeic acid as a green corrosion inhibitor for mild steel," Corrosion Science, vol. 51, no. 3, pp. 642–649, 2008.

84. O. K. Abiola, J. O. E. Otaigbe, and O. J. Kio, "Gossipium hirsutum L. extracts as green corrosion inhibitor for aluminum in NaOH solution," Corrosion Science, vol. 51, no. 8, pp.

1879–1881, 2009.

85. F. Tirbonod and C. Fiaud, "Inhibition of the corrosion of aluminium alloys by organic dyes," Corrosion Science, vol. 18, no. 2, pp. 139–149, 1978.

86. J. D. Talati and J. M. Daraji, "Inhibition of corrosion of B26S aluminium in phosphoric acid by some azo dyes," Journal of the Indian Chemical Society, vol. 68, no. 2, pp. 67–72, 1991.

87. P. Gupta, R. S. Chaudhary, T. K. G. Namboodhiri, B. Prakash, and B. B. Prasad, "Effect of mixed inhibitors on dezincification and corrosion of 63/37 brass in 1% sulfuric acid," Corrosion, vol. 40, no. 1, pp. 33–36, 1984.

88. P. B. Tandel and B. N. Oza, "Performance of some dyestuffs as inhibitors during corrosion of mild-steel in binary acid mixtures (HCl + HNO^3)," Journal of the Electrochemical Society of India, vol. 49, pp. 49–128, 2000.

89. M. L. Zheludkevich, R. Serra, M. F. Montemor, and M. G. S. Ferreira, "Oxide nanoparticle reservoirs for storage and prolonged release of the corrosion inhibitors," Electrochemistry Communications, vol. 7, no. 8, pp. 836–840, 2005.

90. V. V. Torres, R. S. Amado, C. Faia de Sá, et al., "Inhibitory action of aqueous coffee ground extracts on the corrosion of carbon steel in HCl solution," Corrosion Science, vol. 53, no. 7, pp. 2385–2392, 2011.

91. G. G. Geesey, "Microbial exopolymers: ecological and econimic considerations," American Society Microbiology News, vol. 48, pp. 9–14, 1982.

92. I. B. Beech and C. C. Gaylarde, "Recent advances in the study of biocorrosion—an overview," Revista de Microbiologia, vol. 30, no. 3, pp. 177–190, 1999.

93. P. S. Guiamet and S. G. Gomez De Saravia, "Laboratory studies of biocorrosion control using traditional and environmentally friendly biocides: an overview," Latin American Applied Research, vol. 35, no. 4, pp. 295–300, 2005.

94. M. F. Montemor, W. Trabelsi, M. Zheludevich, and M. G. S. Ferreira, "Modification of bis-silane solutions with rare-earth

cations for improved corrosion protection of galvanized steel substrates,"Progress in Organic Coatings, vol. 57, no. 1, pp. 67–77, and 2006.

95. A. Pepe, M. Aparicio, S. Ceré, and A. Durán, "Preparation and characterization of cerium doped silica sol-gel coatings on glass and aluminum substrates," Journal of Non-Crystalline Solids, vol. 348, pp. 162–171, 2004.

96. X. W. Yu, C. N. Cao, Z .M. Yao, Z. Derui, and Y. Zhongda, "Corrosion behavior of rare earth metal (REM) conversion coatings on aluminum alloy LY12," Materials Science and Engineering A, vol. 284, no. 1-2, pp. 56–63, 2000.

97. A. N. Khramov, N. N. Voevodin, V. N. Balbyshev, and M. S. Donley, "Hybrid organo-ceramic corrosion protection coatings with encapsulated organic corrosion inhibitors," Thin Solid Films, vol. 447-448, pp. 549–557, 2004.

98. E. M. Sherif and S. M. Park, "Effects of 1,4-naphthoquinone on aluminum corrosion in 0.50 M sodium chloride solutions," Electrochimica Acta, vol. 51, no. 7, pp. 1313–1321, 2006.

99. M. S. Donley, R. A. Mantz, A. N. Khramov, V. N. Balbyshev, L. S. Kasten, and D. J. Gaspar, "The self-assembled nanophase particle (SNAP) process: a nanoscience approach to coatings," Progress in Organic Coatings, vol. 47, no. 3-4, pp. 401–415, 2003.

100. D. V. Andreeva, D. Fix, H. Möhwald, and D. G. Shchukin, "Self-healing anticorrosion coatings based on pH-sensitive polyelectrolyte/inhibitor sandwichlike nanostructures," Advanced Materials, vol. 20, no. 14, pp. 2789–2794, 2008. ·

101. M. L. Zheludkevich, D. G. Shchukin, K. A. Yasakau, H. Möhwald, and M. G. S. Ferreira, "Anticorrosion coatings with self-healing effect based on nanocontainers impregnated with corrosion inhibitor,"Chemistry of Materials, vol. 19, no. 3, pp. 402–411, 2007.

102. V. DeGiorgi, "Corrosion basics and computer modeling," in Industrial Applications of the BEM, chapter 2, pp. 47–79, 1986.

103. H. Sun, P. Ren, and J. R. Fried, "The compass force field: parameterization and validation for polyphosphazenes," Computational and Theoretical Polymer, vol. 8, pp. 229–246, 1998.

104. N. G. Zamani, J. F. Porter, and A. A. Mufti, "Survey of computational efforts in the field of corrosion engineering," International Journal for Numerical Methods in Engineering, vol. 23, no. 7, pp. 1295–1311, 1986.

105. R. S. Munn, "A review of the development of Computational Corrosion analysis for special corrosion modelling through its maturity in the Mid 1980's-Computer modelling in Corrosion," ASTM STP 1154 American Study for Testing and Materials, Philadelphia, pp 215-228, 1991.

106. K. F. Khaled, "Molecular simulation, quantum chemical calculations and electrochemical studies for inhibition of mild steel by triazoles," Electrochimica Acta, vol. 53, no. 9, pp. 3484–3492, 2008.

107. F. J. Presuel-Moreno, H. Wang, M. A. Jakab, R. G. Kelly, and J. R. Scully, "Computational modeling of active corrosion inhibitor release from an Al-Co-Ce metallic coating," Journal of the Electrochemical Society, vol. 153, no. 11, Article ID 002611JES, pp. B486–B498, 2006.

108. Pradip and B. Rai, "Design of tailor-made surfactants for industrial applications using a molecular modelling approach," Colloids and Surfaces A, vol. 205, no. 1-2, pp. 139–148, 2002.

109. K. F. Khaled and M. A. Amin, "Computational and electrochemical investigation for corrosion inhibition of nickel in molar nitric acid by piperidines," Journal of Applied Electrochemistry, vol. 38, no. 11, pp. 1609–1621, 2008.

110. K. F. Khaled and M. A. Amin, "Electrochemical and molecular dynamics simulation studies on the corrosion inhibition of aluminum in molar hydrochloric acid using some imidazole derivatives," Journal of Applied Electrochemistry, vol. 39, no. 12, pp. 2553–2568, 2009.

Chapter 3

Tantalum Oxide Nanocoatings Prepared by Atomic Layer and Filtered Cathodic Arc Deposition for Corrosion Protection of Steel: Comparative Surface and Electrochemical Analysis

Belén Díaz[a], Jolanta Światowska[a], Vincent Maurice[a], Antoine Seyeux[a], Emma Härkönen[b], Mikko Ritala[b], Sanna Tervakangas[c], Jukka Kolehmainen[c], and Philippe Marcus[a]

[a]Laboratoire de Physico-Chimie des Surfaces (LPCS), Chimie ParisTech - CNRS (UMR 7045), Ecole Nationale Supérieure de Chimie de Paris, 11 rue Pierre et Marie Curie, F-75005 Paris, France

[b]Laboratory of Inorganic Chemistry, University of Helsinki, FIN-00014 Helsinki, Finland

cDIARC-Technology Inc., Kattilalaaksontie 1, 02330 Espoo, Finland

ABSTRACT

A comparative study by Time-of-Flight Secondary Ions Mass Spectrometry and X-ray Photoelectron Spectroscopy, *i–E* polarization curves and Electrochemical Impedance Spectroscopy of the corrosion protection of low alloy steel by 50 nm thick tantalum oxide coatings prepared by low temperature Atomic Layer Deposition (ALD) and Filtered Cathodic Arc Deposition (FCAD) is reported. The data evidence the presence of a spurious oxide layer mostly consisting of iron grown by transient thermal oxidation at the ALD film/substrate interface in the initial stages of deposition and its suppression by pre-treatment in the FCAD process. Carbonaceous contamination (organic and carbidic) resulting from incomplete removal of the organic precursor is the major cause of the poorer sealing properties of the ALD film. No coating dissolution is demonstrated in neutral or acid 0.2 M NaCl solutions. In acid solution localized corrosion by pitting proceeds faster with the ALD than with the FCAD coating. The roles of the pre-existing channel defects exposing the substrate surface and of the spurious interfacial oxide promoting coating breakdown and/or delamination are emphasized.

INTRODUCTION

Coating is one of the most popular methods for corrosion prevention of non passivable metal and alloy substrates. Still the development of highly protective ultra thin films (nanocoatings) of potential application in micro and nano systems or in high precision mechanical systems is a challenging issue. The presence of channel defects connecting substrate and environment through the whole film and limiting the barrier property is a common drawback of many deposition techniques. Chemical Vapour Deposition (CVD) methods offer improved conformality resulting in better sealing than Physical Vapour Deposition (PVD) methods that require rather thick coatings to bury the morphological surface heterogeneities. Recent results have proven that excellent barrier properties can be obtained with oxide nanocoatings deposited on engineering alloys with Atomic Layer Deposition (ALD), a CVD derived method [1], [2], [3] and [4]. The best sealing properties were obtained with pure alumina layers (≤100 nm thick) grown on stainless steel and low alloy steel substrates. A reduction in the corrosion current density reaching four orders of magnitude (depending on thickness, substrate and the specific deposition conditions) could be measured for the coated samples. The effect of plasma-treating the substrate surface prior to the coating was explored more recently showing an improved coating adhesion and further reduction of the coating porosity [5].

The corrosion protection properties of these pure ALD Al_2O_3 layers on low alloy steel substrates were investigated in a chloride containing neutral electrolyte [6]. Unexpectedly the results showed poor chemical stability with the coating dissolving at an average rate of 7 ± 1 nm h^{-1}, consistently measured by electrochemical and surface analysis, as a consequence of the electrolyte penetration through the defective channels of the coating. The oxygen reduction occurring on the substrate surface at the bottom of the defects was proposed to locally increase the pH and provoke the alumina dissolution.

Here we address the corrosion protection provided by tantalum oxide nanocoatings. Tantalum is known to be one of the most corrosion resistant metals [7]. The Pourbaix diagram shows formation of a passive oxide stable over the entire potential-pH range [8]. So tantalum oxide seems to be a good candidate for oxide nanocoatings of high chemical stability. In this study tantalum oxide nanocoatings (50 nm thick) were grown on a low alloy steel substrate by two different methods: a CVD technique (ALD) and a PVD technique (Filtered Cathodic Arc Deposition, FCAD). Besides high conformality ALD is suitable to grow films with precise thickness control and excellent large-area uniformity, accuracy and repeatability [9] and [10]. Its main drawback is a slow growth rate, typically below 0.1 nm per cycle [11], although the processing of large batches in one run can balance this aspect [12]. ALD oxide coatings for corrosion protection were first considered in 1999 on stainless steel [13] and more recently developed on low alloy steel [1], [3], [4] and [6], stainless steel [2], [14], [15] and [16], silver [17], copper [18] and [19] and magnesium alloy [20] substrates. ALD TiO_2 and Al_2O_3 films were also grown on PVD coatings to block pinholes and other defects left in the structure [14] and [21]. Cathodic arc deposition is able to produce high quality hard coatings that offer excellent mechanical properties and suitable corrosion protection [22] and [23]. The generation of macroparticles which limits the range of applications has been successfully addressed by using filtered cathodes [24]. High quality materials and wear resistant films with lower size particle distributions can be currently prepared. The filtered procedure has also been proven to produce FCAD films with improved corrosion resistance [25], [26], [27] and [28].

A comparison of the tantalum oxide layers prepared by these two methods is presented here in terms of chemical architecture, electrochemical behaviour and short-term corrosion protection. Time-of-Flight Secondary Ions Mass Spectrometry (ToF-SIMS) and X-ray Photoelectron Spectroscopy (XPS) were applied in order to characterize the surface, bulk and interfacial chemical composition of the samples. Polarization curves were employed to calculate the uncoated surface fraction or so-called coating porosity and Electrochemical Impedance Spectroscopy (EIS) was employed for analysis of the corrosion prevention during immersion in corrosive sodium chloride aqueous solutions.

EXPERIMENTAL

The substrate material was hardened (805 HV hardness) and tempered (at 180 °C) low alloy steel (DIN 100Cr6, AISI 52100). The composition (wt%) was C (0.95–1.1), Cr (1.5), Ni (max. 0.30), Mn (0.25–0.45), Cu (max. 0.30), Si (0.15–0.35), P (max. 0.030) and S (max. 0.025). The samples were tumble polished to remove oxide layers and other surface residues, ground by planar grinding and lapped with a water-based diamond suspension (6 μm). Prior to coating the substrates were carefully wiped with acetone, ultrasonicated for 5 min in isopropanol, and blow-dried with compressed air.

The ALD process was run in a Picosun SUNALE R-150 reactor. Deionized water (resistivity >18 MΩ cm) and tantalum pentaethoxide $(Ta(OC_2H_5)_5,$ SAFC Hitech™) were used as oxygen and tantalum precursors, respectively. The pulse length was 0.4 s and the purging time was 5 s. The reactor pressure and the inert gas flow were kept at 5 mbar and 600 sccm, respectively. The deposition temperature was 160 °C, which is low for the ALD process [9] but was required by the tempering treatment of the substrate. The deposition rate was 0.04 nm per cycle. The film thickness was measured using a silicon wafer coated simultaneously with the steel substrates, with an UV-VIS spectrophotometer (Hitachi U-2000) and simulated using the Ylilammi and Ranta-aho [29] software.

The FCAD process was carried out in DIARC coating equipment. Samples were pre-etched with 350 eV Ar ions at 0.5 mA cm^{-2} current density for 30 min before coating deposition. The metal oxide coating was produced by depositing Ta plasma from a solid target having minimum 99.8% purity in partial pressure of oxygen. The deposition rate of the metal oxide coating was approximately 0.5 μm h^{-1}. During deposition the samples were heated up to maximum temperature of 50 °C. The coating thickness was controlled from a silicon sample which was fixed in the deposition chamber at a similar position as the steel samples. The thickness was measured with a Dektak 3ST profilometer from the interface between a coated and an uncoated region of the silicon sample after the removal of ink mask. Before analysis all samples were cleaned with ethanol for 10 min in ultrasonic bath and dried with compressed air.

A ToF-SIMS 5 spectrometer (IonTof) operating at a pressure of 10^{-9} mbar was used for elemental depth profiling and chemical mapping. For depth profiling the spectrometer was run in HC-BUNCHED mode with optimum mass resolution (m/ m of 10^4)

but poor lateral space resolution. A pulsed 25 keV Bi$^+$ primary ion source was employed for analysis, delivering ~0.85 pA of current over a 100 μm × 100 μm area. It was interlaced with a 2 keV sputtering Cs$^+$ beam giving a ~80 nA target current over a 400 μm × 400 μm area. The profiles were recorded with negative secondary ions, more sensitive to fragments originating from oxide matrices. For surface chemical mapping, the procedure included a few seconds of sputtering in order to remove the contamination layer existing at the extreme surface. Negative ion chemical maps were then recorded in low primary ion current conditions (non-etching static SIMS conditions) at optimal lateral space resolution (~150 nm) but with poorer mass resolution than for depth profiling. A pulsed 25 keV Bi$^+$ primary ion source was employed for analysis, delivering ~0.2 pA of current over an analyzed area of 100 μm × 100 μm or 300 μm × 300 μm. Data acquisition and post-processing analyses were performed using the Ion-Spec software.

XPS analysis was carried out by using a VG ESCALAB 250 spectrometer operating at a residual pressure of 10^{-9} mbar. An Al K monochromatized radiation (h = 1486.6 eV) was employed as X-ray source. The photoelectrons were collected at a 90° take-off angle with respect to the substrate surface. Survey and high resolution spectra were recorded with pass energies of 100 and 20 eV, respectively. XPS depth profiles were measured using a 1 keV Ar$^+$ sputter beam giving 1 μA of sample current for etching. Survey and high resolution spectra were recorded at selected points of the profile. The data processing was performed with the Avantage® software (Thermo Electron Corporation). Gaussian/Lorentzian (70%/30%) peak shapes were used and Shirley background subtraction was applied. Binding energies were corrected with reference to the C1s signal set at 285 eV. The measured core level intensities were converted into atomic concentrations using Scofield cross-section factors, , and calibrated values for the transmission factor T of the spectrometer.

Electrochemical analysis was performed with a potentiostat/galvanostat Autolab PGSTAT30. A conventional three-electrode cell was employed using a Saturated Calomel Electrode (SCE) and a platinum wire as reference and auxiliary electrodes, respectively. The working electrode area was 0.44 cm². All the experiments were performed at room temperature in solutions prepared with ultra pure water (resistivity >18 MΩ cm) and reagent grade chemicals (NaCl and HCl 37% Analar Normapur analytical reagents, VWR® BDH Prolabo®). The electrolytes were bubbled with Ar for 30 min before test and during experiments.

Potentiodynamic polarization curves were obtained by linear sweep voltammetry (LSV) in a neutral (pH 7) 0.2 M NaCl aqueous solution after an initial period of 1 hour at the open circuit potential (OCP). The samples were polarized from −0.9 V in the anodic direction and the scan was stopped when the current density reached 10 μA cm⁻². The scan rate was 1 mV s⁻¹ for consistency of examination with other oxide nanocoatings deposited on the same steel [1], [3], [4], [6] and [28]. Tafel analysis was performed in order to study the coating efficiency. The polarization resistance (R_p) values provides a reliable assessment of the coating porosity (P, %) [30].

Corrosion tests were conducted at OCP and monitored in situ by impedance measurements in 0.2 M NaCl aqueous solutions at pH 7 and at pH 2 (obtained by adding 0.01 M HCl). The OCP value was recorded for 20 min and then the impedance was performed in the next 10 min to complete the first 30 min of immersion. The procedure was repeated after 1 h of immersion and then every hour for a total immersion period of 6 h. Frequencies between 100 kHz and 10 mHz were used with an amplitude signal set to 10 mV to guarantee a linear response. EIS analysis was complemented by *post mortem* ToF-SIMS analysis.

RESULTS AND DISCUSSION

Coating Characterization

ToF-SIMS Depth Profile Analysis

Fig. 1(A and B) allows to compare the ToF-SIMS depth profiles for the pristine FCAD and ALD layers. The selected ions were $^{12}C^-$, $^{17}OH^-$, $^{18}O^-$, $^{28}Si^-$, $^{35}Cl^-$, $^{52}Cr^-$, $^{56}Fe^-$, $^{84}CrO_2^-$, $^{88}FeO_2^-$, $^{181}Ta^-$, $^{193}TaC^-$ and $^{213}TaO_2^-$. $^{18}O^-$ is the naturally occurring oxygen isotope recorded since the $^{16}O^-$ signal was close to saturation. The ion intensities are presented in logarithmic scale in order to emphasize the low intensity signals, and plotted versus Cs^+ sputtering time. Three regions are distinguished as marked on the profiles. The coating region extends from the beginning of the profile until the position where the intensities of ions characteristic for the coating (TaO_2^-) decrease and those of the ions characteristic for the substrate (Fe^-, Cr^-, Si^- and C^-) increase. It is followed by the interfacial region and then the substrate region, the transition between them being set at the position where the intensity profiles of the coating oxide ions (TaO_2^-) intersect the intensity profiles of ions characteristics of the substrate (Si^-). The same criterion was used for all samples allowing a consistent comparison on the coating architecture.

In the bulk coating region the characteristic TaO_2^- and $^{18}O^-$ ions have steady intensities pointing to homogeneous growth during the deposition process. For the FCAD sample slight oscillations are observed suggesting slight variations of the architecture supported by the XPS data discussed below. In this region the samples differ by a markedly higher organic contamination (C^- and TaC^- ions) of the ALD film.

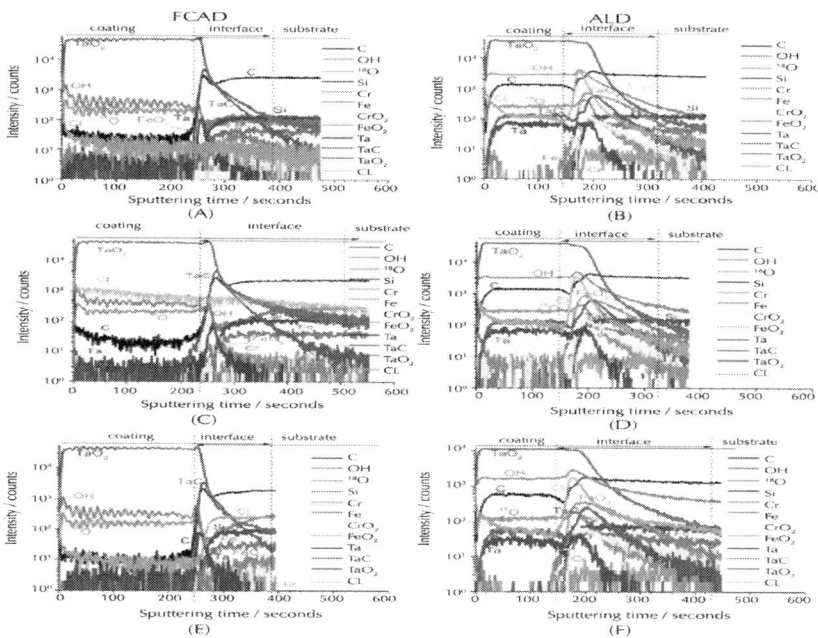

Figure 1: ToF-SIMS negative ions depth profiles for the 50 nm tantalum oxide layer prepared by FCAD (A, C, E) and ALD (B, D, F) on the 100Cr6 substrate before (A, B) and after immersion in neutral (C, D) and acid (E, F) 0.2 M NaCl at OCP for 6 h.

This is consistent with the use of a metal organic precursor for deposition. The use of water as ALD oxygen precursor also results in higher contamination by water residues as shown by the OH^- ion profiles. A significantly shorter time is needed to reach the onset of the interfacial region for the ALD film (146 s) than for the FCAD sample (240 s) showing a higher erosion rate. Thus the low temperature ALD process produces a tantalum oxide film more polluted by precursor residues and possibly less compact [31].

Significant differences between the ALD and FCAD samples are also observed in the interfacial region. The presence of an oxide layer (containing Fe and Cr, Fe:Cr ratio of 8–9) underneath the ALD film is proven by the FeO_2^- and CrO_2^- ions profiles that peak in this region. For the FCAD film only a shallow peak is observed in the profile of the FeO_2^- ions and no peak is detected for the CrO_2^-

ions. The origin of this difference is twofold: (i) an air-formed native oxide is present on the steel surface before deposition and (ii) an interfacial thermal oxide grows transiently during the initial stages of the ALD process due to exposition of the non yet fully covered substrate to water at high temperature (160 °C) [2]. The pre-etching stage in the FCAD process enables to remove the native oxide layer from the substrate surface as discussed previously [28]. Then the coating is grown on a substrate surface that remains almost oxide-free due to deposing first tantalum prior to leaking oxygen gas. The FCAD sample analysis confirms that a duplex Ta/Ta-carbidic layer is developed at the interface as a result of the reaction between the first deposited metallic Ta particles and the residual C contamination as discussed previously [28]. The OH$^-$ and O$^-$ ion intensities in the interfacial region are lower for the FCAD sample, confirming removal of the native oxide and nearly no growth of a thermal oxide. Chlorine contamination of the bulk coating and interfacial regions is also revealed by the Cl$^-$ ion profiles. Trace Cl contamination, brought by the deposition process (possibly the Ta target), is observed to nearly not vary with depth in the FCAD sample. The profile is more intense for the ALD sample layer pointing to higher trace contamination. Some accumulation of these ions is observed at the interface of the ALD film which represents a serious pollution that may affect the service life. A matrix change effect contributing to the different intensities measured between the bulk and the interfacial regions cannot be excluded.

XPS Surface Analysis

Fig. 2 shows the high resolution XP spectra of the C1s, O1s and Ta4f core levels measured at the surface of the FCAD and ALD samples and their fit. Table 1 compiles binding energies (BE), atomic fractions and elemental percentages for each peak component. Like in our preceding work [28] the Ta4f signal was fitted with four $4f_{7/2}/4f_{5/2}$ spin-orbit doublets, 1.9 eV apart in BE and with an intensity ratio set at 4/3 [32] and [33]. The doublet at 26/27.9 eV is assigned to the Ta(V) oxidation state (Ta$_2$O$_5$ oxide matrix) [34], [35],[36] and

[37], and those at lower binding energies are assigned to lower Ta oxidation states (24.8/26.7 eV and 23.9/25.8 eV) and Ta metal (21.7/23.6 eV) [38]. The data show that the FCAD film surface is mostly composed of Ta_2O_5 (82.3%) with a small amount of Ta sub-oxides (15.5%) and non-oxidized Ta (2.2%) (Table 1). In contrast only Ta_2O_5 is identified for the ALD sample. The shoulder at lower binding energies was not observed in this case, which excludes the presence of Ta sub-oxides and Ta metal in noticeable amounts. The near surface region of the ALD film is then more homogeneous in terms of oxide composition than that of the FCAD film.

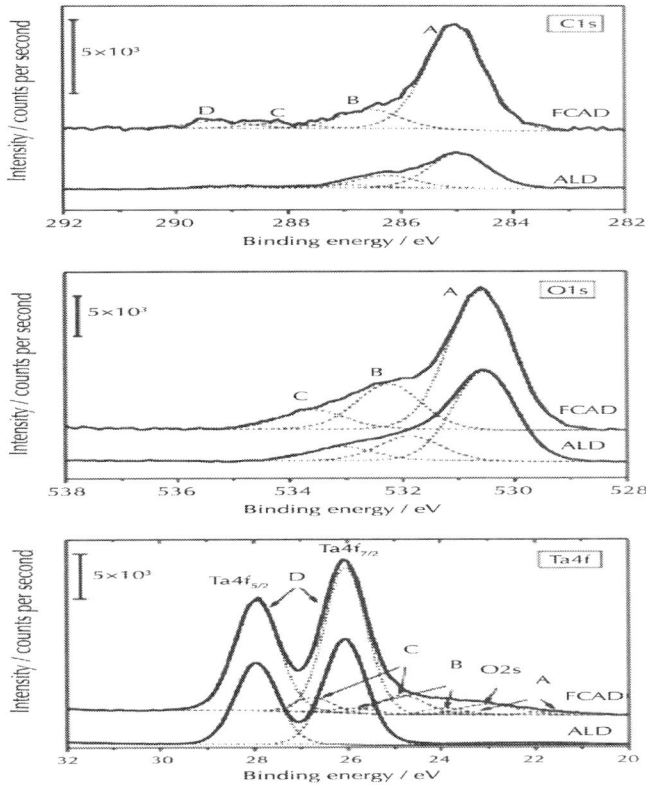

Figure 2: High resolution XP C1s, O1s and Ta4f spectra of the FCAD and ALD film surfaces and their fit. See text and Table 1 for details on peak components.

Table 1: Binding energy values (eV), atomic fraction and elemental percentage of the component peaks of the XP C1s, O1s and Ta4f core levels at the surface of the 50 nm tantalum oxide layers prepared by FCAD and ALD

	FCAD layer			ALD layer		
	BE/eV	Atomic fraction (%)	Relative % of element (%)	BE/eV	Atomic fraction (%)	Relative % of element (%)
C_{1sA}	285	34.4	77	285	21.9	66.6
C_{1sB}	286.6	6.3	14.1	286.3	8.1	24.6
C_{1sC}	288.2	1.6	3.6	287.2	1.4	4.2
C_{1sD}	289.3	2.4	5.3	288.9	1.5	4.6
O_{1sA}	530.6	31.8	68.7	530.6	39.3	69.6
O_{1sB}	532.3	10.1	21.8	531.9	11	19.5
O_{1sC}	533.6	4.4	9.5	533.2	6.2	10.9
$Ta_{4f7/2A}$	21.7	0.2	2.2	–	–	–
$Ta_{4f7/2B}$	23.9	0.3	3.3	–	–	–
$Ta_{4f7/2C}$	24.8	1.1	12.2	–	–	–
$Ta_{4f7/2D}$	26	7.4	82.3	26	10.7	100

Organic contamination of the surface of the films is measured. Four different components are identified as $C1s_A$, $C1s_B$, $C1s_C$ and $C1s_D$ peaks, assigned to C—C, C—O, C=O (carboxyl groups) and O—C=O bonds (carbonates), respectively [3], [39] and [40]. The O1s spectra are decomposed using three different components: the main peak at 530.6 refers to the oxide matrix [41] and [42], the peak at 532.1 ± 0.2 eV can be assigned to hydroxyl groups or contaminants (carboxyls and/or carbonates) [39], [40] and [43] also shown by the C1s spectra and the third peak at 533.4 ± 0.2 eV can be attributed to water molecules adsorbed on the surface [44], [45] and [46]. Table 1 shows that for both samples the major carbon surface contaminant is hydrocarbons originating from air-exposure. The major oxygen surface contaminant is hydroxyl groups originating from water present during deposition but also from air-exposure as indicated by the ToF-SIMS OH⁻ ion profiles that increase at the coating surface. No trace of Cl was detected showing that the intensity measured by ToF-SIMS is below the detection limit of XPS (<0.5 at%).

XPS Depth Profile Analysis

Fig. 3 shows the XPS compositional depth profiles for the FCAD and ALD samples. The data were obtained by analysis of the Ta4f, O1s, Fe2p and C1s core level regions. It can be seen that, like for ToF-SIMS, the XPS depth profiles allow us to distinguish the bulk coating region from the interfacial region. A significant longer time is needed to reach the interface for the FCAD layer since the Fe signal is still fully attenuated by the coating after 3500 s of sputtering whereas it is detected through the coating already after 800 s of sputtering for the ALD sample. This confirms, in line with the ToF-SIMS data, the slower sputtering rate of the FCAD tantalum oxide layer, and thus a different architecture of the pristine coating discussed hereafter.

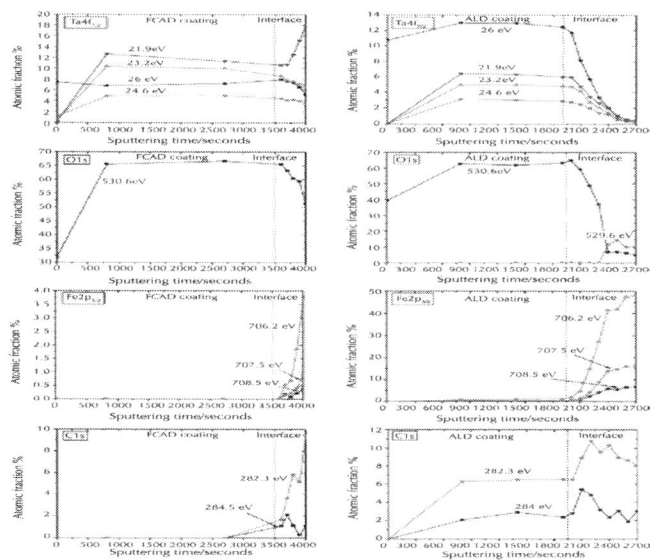

Figure 3: XPS compositional depth profiles for the FCAD (left column) and ALD (right column) samples.

No major BE difference of the Ta4f core level components was observed between the FCAD and ALD samples. In both cases four different peaks were identified. Their intensity variation shows a significant initial increase of the lower BE components after the first sputtering stage. This is (at least partly) associated to oxide reduction by preferential oxygen removal by Ar^+ sputtering [47]. For the FCAD layer the high BE (26 eV) $Ta4f_{7/2}$ component decreases from 82% of the Ta intensity at the coating surface to an average of 20% in the bulk coating region. In parallel the lower BE $Ta4f_{7/2}$ components at 23 ± 0.1 eV and 21.7 ± 0.1 eV show a significant increase from 3.3 % and 2.2% at the surface to 30% and 36%, respectively, in the bulk coating region. The $Ta4f_{7/2}$ peak at 24.6 ± 0.2 eV does not show much variation (from 12.2 to 14%). For the ALD film, the 26 eV component decreases from 100% at the surface to 47% as the sputtering progresses towards the substrate. The lower BE peaks, at 24.6 ± 0.2 eV, 23.1 ± 0.1 eV and 21.7 ± 0.1 eV, show an increase from 0 up to 11, 18 and 24%, respectively. As clearly shown by Fig. 3, while the majority component at the surface of both films corresponds to the Ta(V) oxide matrix (Ta_2O_5), this is the case in the bulk coating region only for the ALD layer. Thus differences between the pristine oxide coating architecture are revealed despite the modifications brought by the reducing sputtering effect. The ALD process allows the development of a Ta(V) oxide layer with no detectable contribution of the tantalum sub-oxides or tantalum metal. In contrast the FCAD process does not permit complete oxidation of the metallic Ta plasma leading to the presence of metallic and sub-oxides tantalum particles in the film. Likely this higher content of metallic tantalum particles in the bulk coating contributes to the slower sputtering rate of the layer. The ToF-SIMS depth profiles did not reveal this difference because negative ions, more suitable for analysis of oxide matrices, were collected.

The O1s core level of the FCAD sample showed both in the bulk coating and interfacial regions only one peak, at 530.7 ± 0.1 eV, characteristic of the Ta—O bonds [32] and [33]. A second peak at lower binding energy, 529.6 ± 0.1 eV, was detected in the interfacial region for the ALD sample. It corresponds to oxygen bonded in another oxide matrix, most likely iron oxide [48] and [49], and thus confirms the presence of a spurious oxide layer in the interfacial region. For the Fe2p core level three Fe2p$_{3/2}$ components located at 706.2 ± 0.1 eV, 707.5 ± 0.2 eV, 708.9 ± 0.4 eV were identified in the interfacial region. They were assigned to metallic Fe, iron carbide and low iron oxidation states, respectively [48], [49], [50] and [51]. The latter one confirms that the spurious interfacial oxide contains iron oxide in much higher concentration in the ALD sample. Low intensity Cr2p core level peaks (not shown) were also detected in the interfacial region of the ALD sample. Two Cr2p$_{3/2}$ components were identified at 573.6 ± 0.1 eV and 575.3 ± 0.3 eV assigned to metallic Cr and chromium oxide, respectively [52] and [53]. The presence of chromium oxide in the spurious interfacial oxide layer of the ALD sample is thus in agreement with the ToF-SIMS data.

Decomposition of the C1s signal showed the existence of two different species. The most intense component located at 282.3 ± 0.2 eV is assigned to carbidic bonds, i.e. either tantalum and/or iron carbides [54] and [55]. The minority component at 284.5 ± 0.2 eV is characteristic of organic contamination. For the FCAD sample, the two components were only detected in the interfacial region. It is worth noting the increase of the metallic Ta component also in the interfacial region. These XPS data corroborate the presence of the Ta/Ta-C duplex layer revealed by ToF-SIMS. The presence of organic contamination at the interface is also shown by XPS. For the ALD sample the two carbidic and organic components are also detected in the bulk coating region. They confirm that the organic precursor used for deposition is not effectively removed from the film due to the low temperature used for the ALD process. The presence of both carbidic and organic residues in non negligible amounts, 6–7 and 2–3 at%, respectively, in the bulk coating is revealed by XPS for this sample. These carbon residues may also contribute in increasing the sputtering rate of the oxide matrix compared to the purer FCAD bulk matrix for which the carbon contamination is below the detection limit of XPS (<0.5 at%).

Coating Porosity

The *i–E* polarization curves obtained in neutral (pH 7) 0.2 M NaCl solution are presented in Fig. 4 for the bare and coated substrates. A markedly reduced current density is measured for both coated samples in the cathodic and anodic branches. There is no significant variation in the shape of the polarization curves for the coated samples. They both show the active behaviour expected from the low Cr content of the alloy. Thus the tantalum oxide layers do not contribute significantly to the electrochemical response, the major effect being the reduction in the current density. This effect is caused by the reduction of the substrate surface exposed to the electrolyte, and the uncoated surface fraction or so-called coating porosity can be calculated by comparing either the corrosion current (i_{corr}) or the polarization resistance (R_p) [1], [2], [3], [4],[5] and [6]. Porosity values (P, %), measured using Eq. (1):

$$P = \frac{R_p^0}{R_p} \cdot 100\%$$

(1)

where R_p^0 and R_p refers to the polarization resistances (determined assuming Tafel behaviour) for the uncoated and the coated samples, respectively, are given in Table 2. A lower corrosion rate and in consequence a lower porosity value is obtained for the FCAD coating. A reduction by a factor of 4 (4.9% vs. 19.4%) is measured compared to the tantalum oxide film of same nominal thickness prepared by ALD.

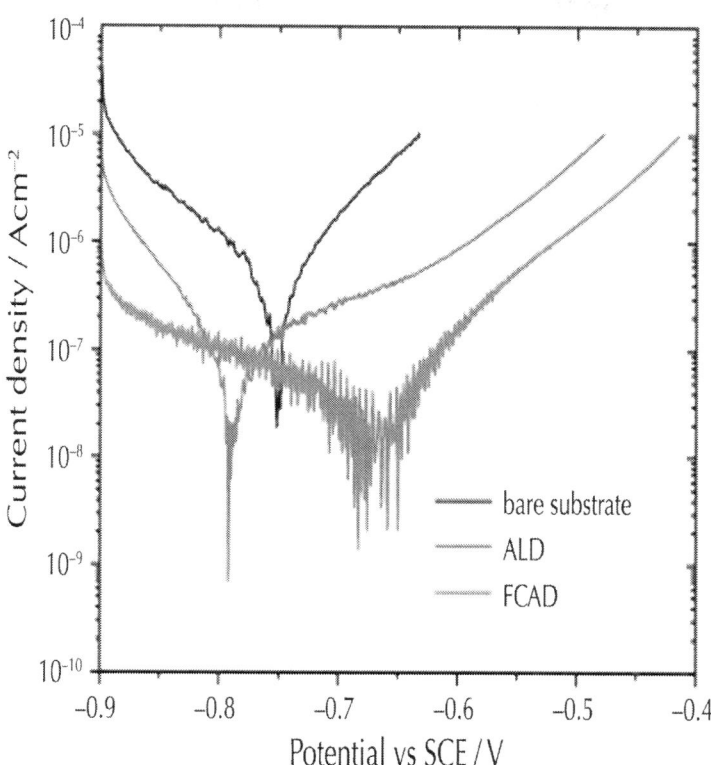

Figure 4: *i–E* polarization curves of the uncoated and coated systems measured in neutral 0.2 M NaCl solution ($dE/dt = 1$ mV s^{-1}).

Table 2: Electrochemical data and porosity values obtained from the polarization curves measured in neutral 0.2 M NaCl

	E_{corr} vs SCE (mV)	$i_{corr} \times 10^{-8}$ (A cm^{-2})	$R_p \times 10^4/\Omega$ cm^2	Porosity (*P*, %)
Bare	−750 ± 5	42.3 ± 0.2	4.8 ± 0.5	100
ALD	−790 ± 5	9.3 ± 0.2	24.7 ± 2	19.4
FCAD	−670 ± 5	3.9 ± 0.2	97.6 ± 10	4.9

The differences on the composition of the interfacial region underneath the coating impact the corrosion potential (E_{corr}).

Compared to the bare substrate, the ALD sample shows a cathodic shift (~-30 mV) while the FCAD sample shows an anodic shift ($\sim+80$ mV). As discussed above, the ALD process induces the formation of a thermal spurious oxide (Fe mostly and Cr oxides), partially hydrated, at the steel surface in the initial stages of deposition process owing to exposure to an oxidizing environment (water at 160 °C). This thermal oxide is expected to subsist at the bottom of the coating pores/defects where the substrate is exposed to the electrolyte. In consequence the corrosion potential shows a more cathodic (less noble) value in agreement with previous data [56]. In contrast, the pre-etching stage included in the FCAD process removes the pre-existing native oxide from the substrate surface resulting in an altered surface state nearly free of iron and chromium oxides after coating. In consequence the corrosion potential shows an anodic shift due to ennoblement.

Thus the presence of native oxides (mostly Fe) on the steel substrate and their transient growth in the initial stages of ALD coating is considered to be detrimental to the corrosion protection if exposed to the aggressive environment by the coating channel defects. However they do not markedly affect the barrier property of the coating as shown by substrate modifications introduced by surface plasma pre-treatments [5]. The poorer sealing performance of the present ALD tantalum oxide nanocoating primarily relates to insufficient removal of the organic metal precursor at the deposition temperature of 160 °C, as shown by the present data and in agreement with previous data on the effect of the deposition temperature [2]. The sealing performance is also very sensitive to the presence of organic pollutants on the substrate surface as previously discussed [3], [4], [5] and [6]. For the FCAD coating, previous work has shown much less beneficial effect of increasing the thickness on the sealing property of FCAD thin films compared to ALD films [28]. This was assigned to the well-known columnar growth of the PVD layers [57] and [58], not enabling efficient sealing of the microstructure defects formed in the initial stages of deposition.

Corrosion Properties

ToF-SIMS Analysis after Immersion

Fig. 1(C and D) and (E and F) shows the depth profiles after immersion of the FCAD and ALD samples in the neutral and acid electrolytes, respectively. For both coated systems no reduction in the sputtering time was measured in the conditions of the test, showing no thickness decrease and thus no generalized dissolution of the coatings. This is in contrast with what was previously observed for Al_2O_3 layers prepared by ALD on the same steel substrate [6]. Thus a good stability in both neutral and acidic electrolyte is demonstrated for these FCAD and ALD tantalum oxide coatings. The oxygen reduction most likely also occurring on the substrate surface at the bottom of the coating defects for the present samples, and locally increasing the pH, does not provoke the oxide dissolution in contrast with the ALD alumina coating. This is assigned to the high stability of tantalum oxide in accordance with the Pourbaix diagram.

After immersion in the neutral electrolyte (Fig. 1(C and D)) no major change was observed in the ion profile intensities indicating no major chemical modification of the coatings. For the ALD sample an increase in the intensities of the FeO_2^- and CrO_2^- ions is observed in the bulk coating region but not for the FCAD sample. This suggests the accumulation of corrosion products at the bottom of the coating defects exposed to the electrolyte. So the coating defects allowing the detection of species at the interface would be larger in the ALD sample than in the FCAD sample, which is consistent with the porosity data reported above. The chemical maps presented in Fig. 5(A and B) also support this conclusion. Defects assigned to pinholes are clearly revealed by the contrast of the TaO_2^- and O^- (not shown) maps. For the FCAD sample (Fig. 5(A)) their sizes range from 5 μm to less than 1 μm while for the ALD film (Fig. 5(B)) larger defects, about 20 μm, are revealed. In both samples, these defects contain OH and Cl. Carbon contamination is also detected in the defects. Oxide layers grown by both ALD and FCAD thus show a defective growth at C-rich sites. The effect on the defect size is less significant for the FCAD film due to pre-etching prior to deposition. Moreover residues of the organic precursor used in ALD may accumulate on the metallic substrate promoting local defective growth of the coating.

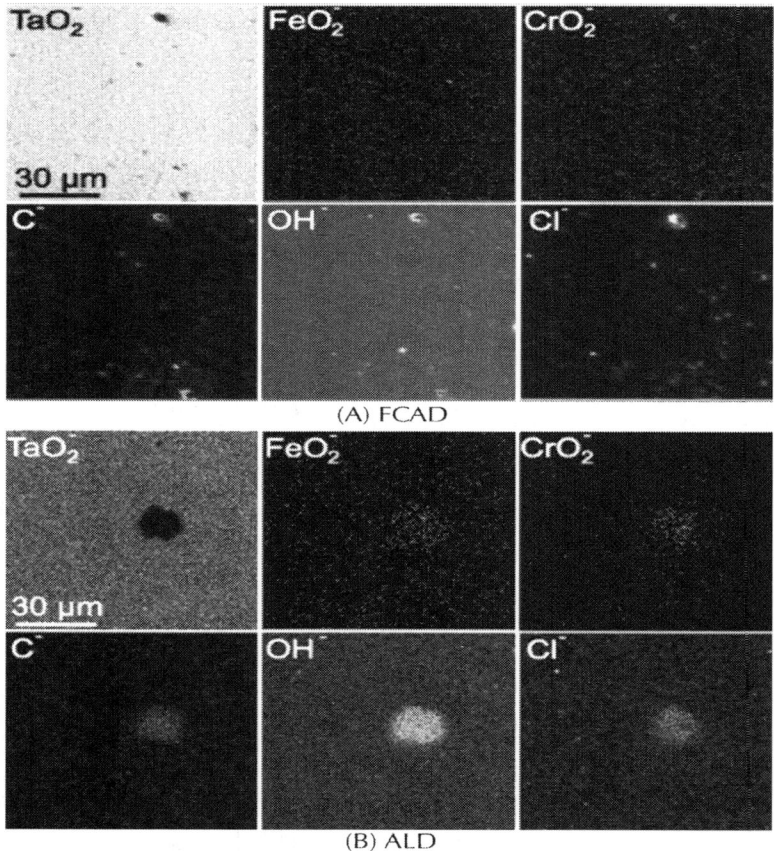

Figure 5: ToF-SIMS negative ion chemical maps for the FCAD (A) and ALD (B) tantalum oxide layer following immersion in neutral (pH 7) 0.2 M NaCl at OCP for 6 h.

After immersion in the acid electrolyte, the depth profiles presented in Fig. 1(E and F) do not show any significant increase of FeO_2^- and CrO_2^- ion intensities in the bulk coating regions, in particular for the ALD sample. This is consistent with dissolution of the corrosion products at low pH [59]. The interfacial regions appear slightly modified. A moderate increase in the OH^- peak intensity is measured in both coated samples, indicating the accumulation of aqueous corrosion products at the interfaces. For the FCAD sample Cl accumulation is observed at the interface showing the penetrating attack of the substrate by pits. For the ALD sample Cl accumulation is also noticeable at the end of the interfacial region. For this sample the longer time needed to sputter the interface region after immersion relates to the roughness increase caused by the corrosion process developed at the bottom of the coating defects and presumably affecting a larger surface fraction. The chemical maps presented in Fig. 6(A and B) illustrate the stronger development of localized corrosion in this more aggressive conditions, much more marked for the ALD sample in agreement with its higher coating porosity. For this sample (Fig. 6(B)), corrosion has invaded the surface with the formation of large rounded pits (10–100 μm wide) and filaments (~5 μm across). Pits, well-marked in the TaO_2^- map, contain Fe corrosion products, as well as Cl, OH and C. Cr corrosion products are also detected mostly at the periphery. Filaments result from anisotropic corrosion, likely initiated at the coating pre-existing defects and propagating by breakdown and/or delamination of the coating during immersion. They also contain Fe, Cl, OH and C but no Cr. The contrast is less marked than in the large pits, suggesting that they correspond to sites of less advanced localized corrosion. For the FCAD sample (Fig. 6(A)), a much more efficient corrosion protection is verified. The pits are smaller (10–15 μm), no filament is formed and a lower surface fraction is corroded. The pits are marked by corrosion products containing Fe and Cr as well as C, OH and Cl contaminations, as for the ALD sample.

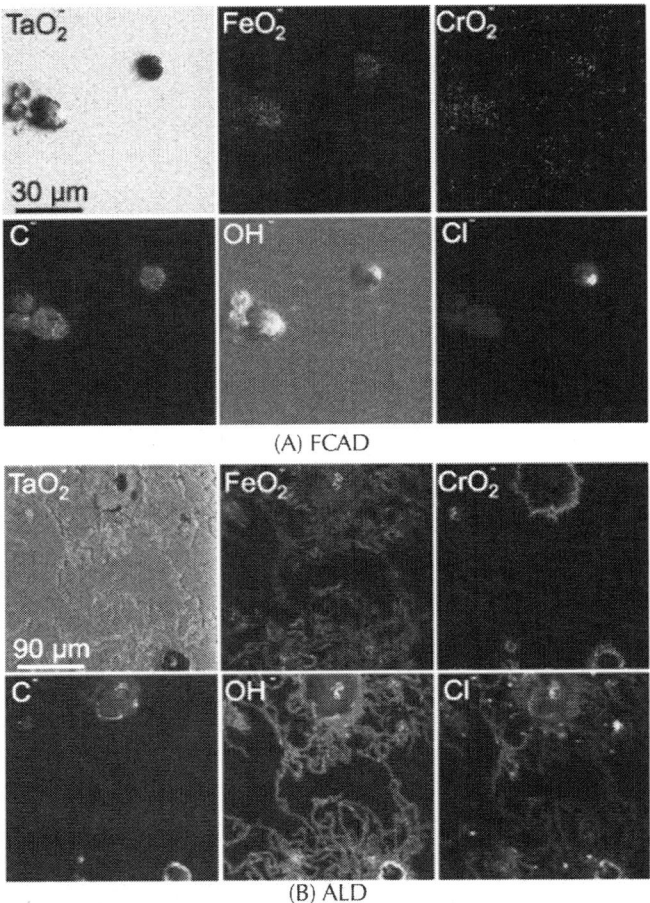

Figure 6: ToF-SIMS negative ion chemical maps for the FCAD (A) and ALD (B) tantalum oxide layer following immersion in acid (pH 2) 0.2 M NaCl at OCP for 6 h.

EIS Analysis during Immersion

Fig. 7 shows the impedance spectra (Nyquist plots) for the FCAD and ALD samples recorded after 0.5, 1 and 6 h in the neutral 0.2 M NaCl solution at OCP. The FCAD sample (Fig. 7(A)) shows a larger global impedance than the ALD sample (Fig. 7(B)) in agreement with the lower uncoated surface fraction (i.e. lower porosity) measured by polarization and discussed above. For this sample no pronounced variation of the global impedance was observed after immersion. This confirms good stability, in agreement with the minor modifications detected by ToF-SIMS.

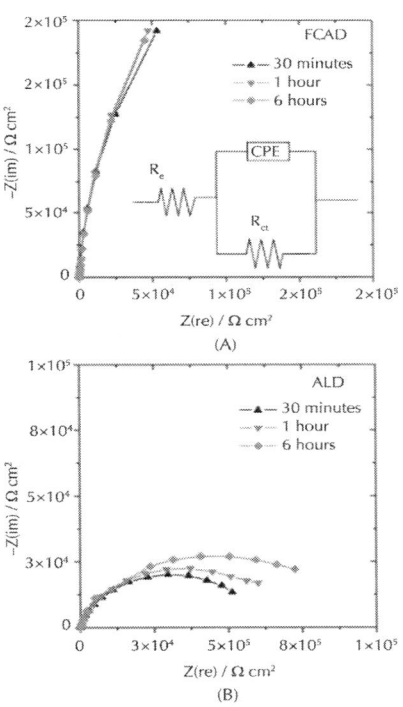

Figure 7: Time evolution of the impedance spectra (Nyquist plots) for the FCAD (A) and ALD (B) coated samples during immersion in neutral 0.2 M NaCl at OCP. The Randles circuit used for data modelling is inserted in (A).

For the ALD sample a slight increase of the global impedance is observed during immersion, in agreement with the more pronounced modifications observed by ToF-SIMS. The accumulation of iron and chromium corrosion products at the bottom of the coating defects exposed to the electrolyte was deduced from the ToF-SIMS data, so that the presence of a more electronically resistive film at the bottom of the defects exposed to the electrolyte can be expected. The OCP variation (Table 3) is consistent with this statement. A cathodic shift is observed during immersion supporting the formation of a less conductive film at the interface [56]. The FCAD sample also showed a cathodic shift of the OCP values during immersion consistent with the accumulation of corrosion products at the bottom of the coating defects. However, this accumulation seems insufficient to be detected in the ToF-SIMS depth profiles, most likely owing to the reduced surface fraction exposed to the electrolyte and to smaller defects exposing the substrate.

Table 3: Best fitting parameters obtained using the Randles circuit displayed in Fig.7(A) for the impedance spectra measured during immersion in neutral electrolyte

	Immersion time/hours	E_{corr} vs SCE (mV)	$R_{ct} \times 10^4$ (Ω cm^2)	$C \times 10^{-7}$ (F cm^{-2})	CPE$_{power}$ (n)
ALD	0.5	-730 ± 5	5.4 ± 0.5	8.1 ± 0.2	0.84 ± 0.01
	1	-740 ± 5	6.1 ± 0.5	8.4 ± 0.2	0.84 ± 0.01
	3	-760 ± 5	6.7 ± 0.5	9.9 ± 0.2	0.83 ± 0.01
	6	-760 ± 5	7.1 ± 0.5	12.9 ± 0.2	0.87 ± 0.01
FCAD	0.5	-630 ± 5	124.8 ± 2	85.1 ± 0.5	0.96 ± 0.01
	1	-650 ± 5	136.9 ± 2	84.3 ± 0.5	0.95 ± 0.01
	3	-670 ± 5	141.7 ± 2	84.1 ± 0.5	0.94 ± 0.01
	6	-670 ± 5	144.1 ± 2	81.5 ± 0.5	0.95 ± 0.01

The Randles circuit, inserted in Fig. 7(A), was used for data fitting. It accounts for the electrolyte resistance, R_e, in the high frequency limit, for the charge transfer resistance, R_{ct}, inversely proportional to the corrosion rate and for a constant phase element (CPE) describing the capacitive contribution and its deviation from the ideal behaviour due to surface heterogeneities such as roughness, scratches, absorption of ions or possible variations in the physical properties of the covering film (resistivity, permittivity) [60]. Capacitance values can be precisely calculated from the CPE assuming the Brug approach [61] and [62]. Two parallel contributions must then be taken into account: the double layer capacitance, C_{dl}, at the electrolyte/uncoated interface, and the coating capacitance, C_{coat}. Unlike for compact films of remarkably low porosities (below ~0.1 %), for which the measured capacitance nearly entirely refers to C_{coat}, the C_{dl} component must also be considered for the present films [3].

Table 3 presents the fitted values of the parameters. For the ALD sample the increase of the global capacitance reflects an increase in C_{dl}. Indeed, since a decrease of the coating thickness can be excluded, as proven by the ToF-SIMS depth profiles, the only possibility to explain the increase of the global capacitance is assuming an increase in the area exposed to the electrolyte. This would imply coating breakdown and/or pit growth in agreement with the ToF-SIMS data. For the FCAD sample, the variation in the capacitance is barely detected, in agreement with the higher stability of this system. One also notices that the capacitance values are one order of magnitude higher for the FCAD film. As previously discussed [28], a dielectric constant higher than theoretically expected would explain this high value. Ta metallic particles, detected by XPS analysis, dispersed in the bulk coating could contribute as parallel elements to the global capacitance thus giving the non expected high dielectric constant [63].

Fig. 8 shows the impedance spectra (Nyquist and Bode plots) for the FCAD and ALD samples immersed 0.5, 1, 2, 3, 4, 5 and 6 h in the more aggressive acid 0.2 M NaCl solution at OCP. A remarkably different behaviour is observed in this electrolyte characterized by a marked reduction of the global impedance due to localized corrosion. A higher corrosion rate than in the neutral electrolyte is even observed already after 0.5 h. At this stage, the total resistance values are already one and two orders of magnitude lower than those measured for the FCAD and ALD systems, respectively, in the mild neutral solution after 6 h of immersion.

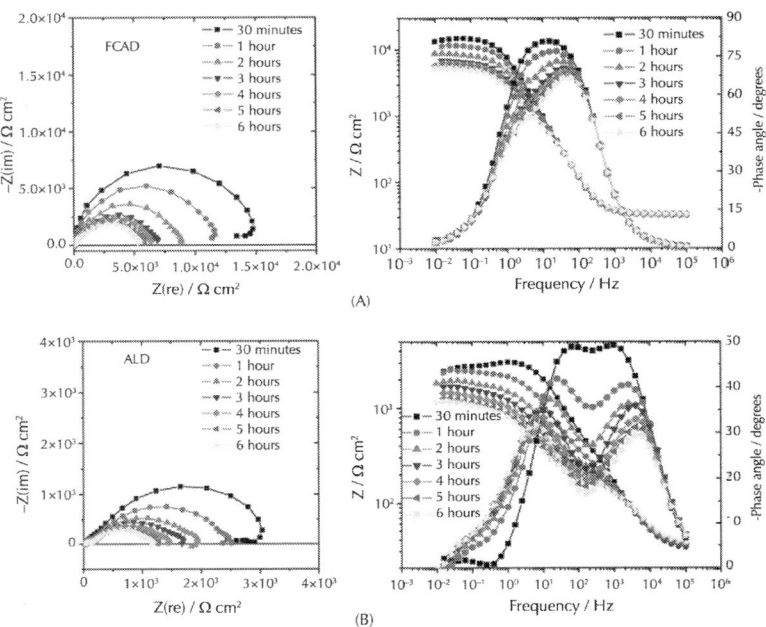

Figure 8: Time evolution of the impedance spectra (Nyquist and Bode plots) for the FCAD (A) and ALD (B) coated samples during immersion in acid 0.2 M NaCl at OCP.

For both coated systems three semicircles, more clearly resolved in the ALD case, are distinguished: two capacitive semicircles at high and medium frequency and an inductive loop at low frequency. This low-frequency loop is changed to another capacitive semicircle as immersion prolongs. Such transition is widely referred to as due to an increase in the surface coverage of intermediate species formed during steel dissolution in acidic media [64], [65] and [66]. When the coverage of these intermediate species is low the impedance diagrams are characterized by a low-frequency inductive loop. When their coverage increases capacitive semicircles are developed in the low frequency domain [67], [68], [69] and [70].

The equivalent circuits used for data fitting are shown in Fig. 9. They are adaptations of previous studies to our systems [64], [66], [67] and [71]. Two different models differing in the low frequency domain were considered: one with the inductive loop, describing the initial stages of the process (Fig. 9, top left diagram), and one with the capacitive semicircle, describing more advanced localized corrosion in the later stages (Fig. 9, bottom left diagram). Several processes are considered in these models: (i) the response of the coated surface (CPE_{coat}, coating capacitance) is arranged in parallel with the response of the uncoated surface (R_{ct}, charge transfer resistance), (ii) the impedance of the pits is included in parallel using in series disposition of the pits' resistance (R_{pit}, electrolyte resistance inside the pit) and double layer capacitance at the substrate/electrolyte interface (CPE_{dl}, including both the pitted and un-pitted surface), (iii) the low frequency loop is considered as an in series arrangement of the adsorbed intermediates' resistance (R_{ads}) and an inductive element (L) at low coverage of the intermediates or as a parallel arrangement of the adsorbed intermediates' resistance (R_{ads}) and the adsorbed intermediates' capacitance (CPE_{ads}) at high coverage. CPEs were again considered instead of pure capacitances.

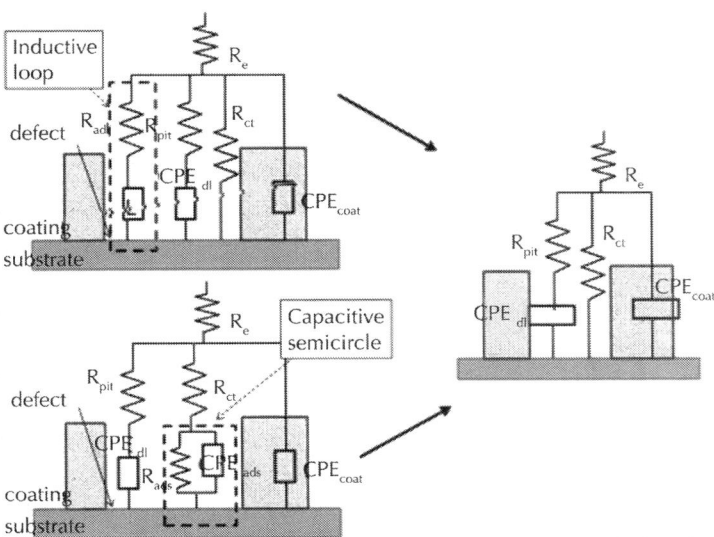

Figure 9: Equivalent circuit used for modelling the impedance spectra in Fig. 8, see text for details

Table 4: Best fitting parameters obtained using the equivalent circuit displayed in Fig.9 for the impedance spectra measured during immersion in acidic electrolyte

	Immersion time/ hours	E_{corr} vs SCE (mV)	R_{ct} (Ω cm²)	$C_{coat} \times 10^{-7}$ (F cm^{-2})	CPE_{power} (n)	R_{pit} (Ω cm²)	$C_{dl} \times 10^{-7}$ (F cm^{-2})	CPE_{power} (n)
ALD	0.5	-570 ± 5	3243 ± 10	5.9 ± 0.5	0.78 ± 0.01	764 ± 5	1.6 ± 0.2	0.84 ± 0.01
	1	-590 ± 5	2435 ± 10	5.4 ± 0.5	0.78 ± 0.01	310 ± 5	4.1 ± 0.2	0.74 ± 0.01
	3	-620 ± 5	1521 ± 10	4.9 ± 0.5	0.78 ± 0.01	225 ± 5	16.4 ± 0.5	0.74 ± 0.01
	6	-620 ± 5	1076 ± 10	4.8 ± 0.5	0.74 ± 0.01	176 ± 5	36.8 ± 0.5	0.72 ± 0.01
FCAD	0.5	-560 ± 5	$15{,}045 \pm 50$	134 ± 5	0.95 ± 0.01	–	–	–
	1	-580 ± 5	$11{,}569 \pm 50$	130 ± 5	0.92 ± 0.01	–	–	–
	3	-600 ± 5	6933 ± 10	128 ± 5	0.93 ± 0.01	2447 ± 10	25.2 ± 0.5	0.77 ± 0.01
	6	-610 ± 5	5467 ± 10	125 ± 5	0.93 ± 0.01	1665 ± 10	49.3 ± 0.5	0.76 ± 0.01

Table 4 compiles the values obtained after fitting. Capacitance values are presented after transformation of the CPEs using the Brug's approach. The data of the low frequency loop were not considered for the fittings since their discussion would be beyond the aim of this paper (see simplified model in Fig. 9, right diagram). It is however worth noting the slower and less pronounced evolvement from inductance to capacitance for the FCAD sample (Fig. 8(A), Nyquist and Bode plots). This difference is a first indication of a slower development of localized corrosion for this sample in this electrolyte. One can also note that the time constant in the medium frequency domain (10^{-1} Hz) is not well resolved for the FCAD sample before the second hour of immersion, confirming a slower kinetics of pitting than for the ALD sample for which both semicircles are already distinguished after 0.5 h of immersion (Fig. 8, Bode plots).

Pitting corrosion is also demonstrated by considering separately the variations of the capacitance and resistance values. The coating capacitance (C_{coat}) can be described by Eq. (2):

$$C_{coat} = \frac{\varepsilon \varepsilon_0 S_{covered}}{d_{coat}}$$

(2)

in which ε refers to the dielectric constant of the layer, ε_0 corresponds to the vacuum dielectric constant ($8.85 \ 10^{-14}$ F cm^{-1}), d_{coat} is the coating thickness and $S_{covered}$ is the coated surface. The C_{coat} decrease during immersion (Table 4) can be accounted by the decrease of the surface, $S_{covered}$, covered by the coating and thus implies coating breakdown. The concomitant increase of the double layer capacitance (C_{dl}) shown in Table 4 agrees also with the enlargement of the uncovered area. This increase of the uncoated surface is assigned to both the development of new pits (implying coating breakdown) and the growth (in width and/or depth) of the existing pits.

Increasing corrosion rate during immersion is also proven by the decrease of the charge transfer resistance R_{ct}. The reduction in the semicircle at the high frequency domain (R_{pit}) completely agrees with the creation of new pits (the pitting resistance includes the contribution of a number pits located in parallel) and/or the growth of larger pits. The electrolyte resistance inside the pits is smaller for larger pits since resistance is inversely proportional to the pitted area.

The coating capacitance values obtained for the ALD film are as expected (4.8×10^{-7} F cm^{-2}) assuming an average value of 26 ± 1 for the dielectric constant [72] and [73]. However, the calculation of the exact covered surface fraction (needed to evaluate precisely C_{coat}) becomes a complicated task as pitting progresses and then the obtained value for the coating capacitance may be inaccurate. Still the measurement shows a variation of -20% of the capacitance. The coating capacitance values for the FCAD layer are larger due to the contribution of the Ta metallic particles as discussed above. The variation, about -7%, is less pronounced for the same immersion period. This result agrees with the less severe attack demonstrated by the ToF-SIMS chemical maps than for the ALD sample.

The evolution measured for the other parameters is also consistent with a lower corrosion of the FCAD sample. Although the evolution of C_{dl} is similar for both samples, the values obtained at the end of the corrosion test are lower for the FCAD sample. This difference suggests that a longer immersion period would be needed to reach an equivalent attack for the more resistant FCAD sample. The pitting resistance after 6 h of immersion is also larger by one order of magnitude for this sample indicating less advanced localized corrosion in agreement with the ToF-SIMS chemical maps. The increase in the surface heterogeneity caused by pitting can be also followed by comparing the evolution in the CPE power (n) values. Here again the variation during immersion is more pronounced for the ALD sample, in particular that for the CPE assigned to the second time constant related to the pit area. In contrast no variation of this parameter is observed for the FCAD sample [3], [74] and [75], confirming higher pitting resistance.

Thus both EIS analysis during immersion and ToF-SIMS analysis after immersion allow us to conclude on pitting corrosion of the coated samples, markedly faster in the more aggressive acid electrolyte. The pits presumably initiate at the pre-existing channel defects of the coating exposing the substrate surface to the electrolyte. The larger uncoated surface fraction is thus one key parameter for the observed more severe attack of the ALD sample. However it is most likely that coating breakdown also occurs and promotes pit initiation. The spurious oxide layer of mostly Fe found in the interfacial region of the coated samples plays here a key role. This layer, unstable in the chloride corrosive environment if not enriched in Cr as for the present samples, will dissolve not only at the bottom of the defects but also along the coating/substrate interface, thus trenching the interface and causing interfacial voiding. As a result coating breakdown and/or delamination will be promoted. We propose that such a mechanism is also at the origin of the much lower pitting corrosion resistance of the ALD sample for which the spurious growth of a mostly Fe oxide was previously evidenced [1], [3], [4] and [6] and further demonstrated by the present data. Pre-etching in the FCAD process is effective in limiting to traces this interfacial oxide layer as shown previously [28] and confirmed here. This is a key parameter to optimize the corrosion protection of steel substrates by oxide nanocoatings.

CONCLUSIONS

ToF-SIMS and XPS depth profile analysis have been combined with voltammetry and EIS electrochemical analysis to study the corrosion protection of low alloy steel by tantalum oxide nanocoatings (50 nm thick) grown by FCAD and by low temperature (160 °C) ALD.

The ToF-SIMS and XPS data show that pre-etching by ion bombardment prior to actual oxide growth in the FCAD process permits to reduce the native oxide layer formed on the steel surface to traces of iron oxide whereas a spurious thermal oxide layer of Fe (mostly) and Cr is grown in the initial stages of ALD due to exposure of the substrate to water vapour at 160 °C. A carbidic interlayer at the interface of the FCAD sample is formed from the reaction between the first deposited metallic particles and the residual carbon contamination of the substrate surface left by pre-etching.

The chemical architectures of the bulk coatings show no significant in-depth variation but are different. Ta(V) oxide (Ta_2O_5) is the major component obtained in both processes but minor Ta sub-oxides and Ta metal are present due to incomplete oxidation of the Ta metal plasma in the FCAD process, whereas only Ta_2O_5 is formed in the ALD process. Bulk hydroxyl, organic and carbidic contamination is present in both coatings but is more pronounced for the ALD sample owing to incomplete removal of the water and organic precursors. The C contamination, reaching 8–10 at% as measured by XPS, is the major cause of the poorer sealing properties of the ALD film grown at 160 °C.

ToF-SIMS combined with EIS analysis performed at increasing immersion time in neutral and acidic NaCl solutions offers a reliable assessment of the corrosion protection mechanism. No coating dissolution was detected. The immersion tests in neutral solution did not reveal significant coating degradation, except minor coating breakdown and/or pit growth for the ALD sample. In the more aggressive acid solution (pH 2) the resistance to localized corrosion is lower and pitting proceeds faster with the ALD coating. Both EIS analysis during immersion and ToF-SIMS analysis after immersion allow concluding to faster coating breakdown and pit growth. The pits presumably initiate at the pre-existing channel defects of the coating exposing the substrate surface to the electrolyte, emphasizing the role of the uncoated surface fraction. It is also proposed that the spurious interfacial oxide underneath the ALD coating is detrimental to corrosion protection, promoting interfacial voiding and coating breakdown and/or delamination by dissolution in aggressive environment.

ACKNOWLEDGMENTS

The research leading to these results has received funding from the European Community's Seventh Framework Programme (FP7/2007-2013) under grant agreement no. CP-FP 213996-1 (CORRAL). Region Ile–de–France is acknowledged for partial support for the ToF–SIMS equipment.

REFERENCES

1. S.E. Potts, L. Schmalz, M. Fenker, B. Díaz, J. Swiatowska, ´V. Maurice, A. Seyeux, P. Marcus, G. Radnóczi, L. Toth, M.C.M. van de Sanden, W.M.M. Kessels, Ultra-thin aluminium oxide films deposited by plasma-enhanced atomic layer deposition for corrosion protection, Journal of the Electrochemical Society 158 (2011) C132.

2. B. Díaz,J. Swiatowska, ´ V.Maurice,A. Seyeux,B.Normand, E.Härkönen,M.Ritala, P. Marcus, Electrochemical and ToF-SIMS analysis of ultra-thin metal oxide (Al2O3 and Ta2O5) coatings deposited by atomic layer deposition on stainless steel, Electrochimica Acta 56 (2011) 10516.

3. B. Díaz, E. Härkönen, J. Swiatowska, ´V. Maurice, A. Seyeux, P. Marcus, M. Ritala, Low-temperature atomic layer deposition of Al2O3 thin coatings for corrosion protection of steel: surface and electrochemical analysis, Corrosion Science 53 (2011) 2168.

4. E. Härkönen, B. Díaz, J. Swiatowska, ´V. Maurice, A. Seyeux, M. Vehkamäki, T. Sajavaara, M. Fenker, P. Marcus, M. Ritala, Corrosion protection of steel with oxide nanolaminates grown by atomic layer deposition, Journal of the Electrochemical Society 158 (2011) C369.

5. E. Härkönen, S. E. Potts, B. Díaz, M., Kariniemi, A., Seyeux, J. Swiatowska, ´ V., Maurice, P., Marcus, G. Radnóczi, L. Tóth, W. M. M. Kessels, M. Ritala, Hydrogenargon plasma pre-treatmentfor improving the anti-corrosion properties ofthin Al2O3 films on carbon steel deposited using atomic layer deposition, Thin Solid Films, submitted for publication.

6. B. Díaz, E. Härkönen, V. Maurice, J. Swiatowska, ´A. Seyeux, M. Ritala, P. Marcus, Failure mechanism of thin Al2O3 coatings grown by atomic layer deposition for corrosion protection of carbon steel, Electrochimica Acta 56 (2011) 9609.

7. E. McCafferty, Introduction to Corrosion Science, Springer, New York, 2010.

8.   M. Pourbaix, Atlas of Electrochemical Equilibria in Aqueous Solutions, National Association of Corrosion Engineers, Houston, TX, 1974.

9.   M. Ritala, M. Leskela, Atomic Layer Deposition, in: H.S. Natwa (Ed.), Handbook of thin film materials, vol. 1, Academic Press, San Diego, 2002, p. 103 (Ch. 2).

10.   S.M. George, Atomic layer deposition: an overview, Chemical Reviews 110 (2010) 111.

11.   G. Cao, Nanostructures and Nanomaterials, Synthesis, Properties and Applications, Imperial College Press, London, 2004.

12.   M. Leskela, M. Ritala, Atomic layer deposition chemistry: recent developments and future challenges, Angewandte Chemie International Edition 42 (2003) 5548.

13.   R. Matero, M. Ritala, M. Leskelä, T. Salo, J. Aromaa, O. Forsén, Atomic layer deposited thin films for corrosion protection, Journal de Physique IV France 9 (1999) 493.

14.   C.X. Shan, X. Hou, K.L. Choy, P. Choquet, Choquet, Improvement in corrosion resistance of CrN coated stainless steel by conformal TiO2 deposition, Surface and Coatings Technology 202 (2008) 2147.

15.   C.X. Shan, X. Hou, K.L. Choy, Corrosion resistance of TiO2 films grown on stainless steel by atomic layer deposition, Surface and Coatings Technology 202 (2008) 2399.

16.   E. Marin, A. Lanzutti, L. Guzman, L. Fedrizzi, Corrosion protection of AISI 316 stainless steel by ALD alumina/titania nanometric coatings, Journal of Coatings Technology and Research 8 (2011) 655.

17.   L. Paussa, L. Guzman, E. Marin, N. Isomaki, L. Fedrizzi, Protection of silver surfaces against tarnishing by means of alumina/titania-nanolayers, Surface and Coatings Technology 206 (2011) 976.

18.   A.I. Abdulagatov, Y. Yan, J.R. Cooper, Y. Zhang, Z.M. Gibbs, A.S. Cavanagh, R.G. Yang, Y.C. Lee, S.M. George, Al2O3 and TiO2 atomic layer deposition on copper for water corrosion

resistance, ACS Applied Materials & Interfaces 3 (2011) 4593.

19. .M.L. Chang, T.C. Cheng, M.C. Lin, H.C. Lin, M.J. Chen, Improvement of oxidation resistance of copper by atomic layer deposition, Applied Surface Science 258 (2012) 10128.

20. E. Marin, A. Lanzutti, L. Guzman, L. Fedrizzi, Chemical and electrochemical characterization of TiO2/Al2O3 atomic layer depositions on AZ-31 magnesium alloy, Journal of Coatings Technology and Research 9 (2012) 347.

21. E. Marin, L. Guzman, A. Lanzutti, L. Fedrizzi, M. Saikkonen, Chemical and electrochemical characterization of hybrid PVD + ALD hard coatings on tool steel, Electrochemistry Communications 11 (2009) 2060.

22. V.I. Gorokhovskya, D.G. Bhatb, R. Shivpuric, K. Kulkarnic, R. Bhattacharyad, A.K. Raid, Characterization of large area filtered arc deposition technology: part II — coating properties and applications, Surface and Coatings Technology 140 (2001) 215.

23. B.K. Tay, Z.W. Zhao, D.H.C. Chua, Review of metal oxide films deposited by filtered cathodic vacuum arc technique, Materials Science and Engineering R52 (2006) 1.

24. D.Y. Wang, C.L. Chang, K.W. Wong, Y.W. Li, W.Y. Ho, Improvement of the interfacial integrity of (Ti,Al)N hard coatings deposited on high speed steel cutting tools, Surface and Coatings Technology 120–121 (1999) 394.

25. P.J. Martin, A. Bendavid, R.P. Netterfield, T.J. Kinder, F. Jahan, G. Smith, Plasma deposition oftribological and opticalthin film materials with a filtered cathodic arc source, Surface and Coatings Technology 112 (1999) 257.

26. C.H. Hsu, C.Y. Lee, C.C. Lee, Analysis on the corrosion behavior of DC53 tool steel coated by Ti–Al–C–N films via filtered cathodic arc deposition, Thin Solid Films 517 (2009) 5212.

27. C.H. Hsu, C.C. Lee, W.Y. Ho, Filter effects on the wear and corrosion behaviors of arc deposited (Ti,Al)N coatings for application on cold-work tool steel, Thin Solid Films 516

(2008) 4826.

28.   B. Díaz, J. Swiatowska, ´V. Maurice, M. Pisarek, A. Seyeux, S. Zanna, S. Tervakangas, J. Kohlemainen, P. Marcus, Chromium and tantalum oxide nanocoatings prepared by Filtered Cathodic Arc Deposition for corrosion protection of carbon steel, Surface and Coatings Technology 206 (2012) 3903.

29.   M. Ylilammi, T. Ranta-aho, Optical determination of the film thicknesses in multilayer thin film structures, Thin Solid Films 232 (1993) 56.

30.   W. Tato, D. Landolt, Electrochemical determination of the porosity of single and duplex PVD coatings of titanium and titanium nitride on brass, Journal of the Electrochemical Society 145 (1998) 4173.

31.   D. Alamarguy, J.E. Castle, N. Ibris, A.M. Salvi, Factors influencing charge capacity of vanadium pentoxide thin films during lithium ion intercalation/deintercalation cycles, Journal of Vacuum Science & Technology A 25 (2007) 1577.

32.   O. Kerrec, D. Devilliers, H. Groult, P. Marcus, Study of dry and electrogenerated Ta_2O_5 and $Ta/Ta_2O_5/Pt$ structures by XPS, Materials Science and Engineering B 55 (1998) 134.

33.   E. Atanassova, G. Tyuliev, A. Paskaleva, D. Spassov, K. Kostov, XPS study of N_2 annealing effect on thermal Ta_2O_5 layers on Si, Applied Surface Science 225 (2004) 86.

34.   Y.Imai,A.Watanabe,M.Mukaida,K. Osato, T. Tsunoda, T.Kameyama,K. Fukuda, Stoichiometry of tantalum oxide films prepared by KrF excimer laser-induced chemical vapor deposition, Thin Solid Films 261 (1995) 76.

35.   F. Gitmans, Z. Sitar, P. Günter, Growth of tantalum oxide and lithium tantalate thin films by molecular beam epitaxy, Vacuum 46 (1995) 939.

36.   J.Y. Zang, I.W. Boyd, Formation of silicon dioxide layers during UV annealing of tantalum pentoxide film, Applied Surface Science 168 (2000) 234.

37.   S. Lecuyer, A. Quemerais, G. Jezequel, Composition of natural oxide films on polycrystalline tantalum using XPS electron

take-off angle experiments, Surface and Interface Analysis 18 (1992) 257.

38. C.H. An, K. Sugimoto, Ellipsometric examination of structure and growth rate of metallorganic chemical vapor deposited Ta2O5 films on Si(100), Journal of the Electrochemical Society 141 (1994) 853.

39. G. Beamson, D. Briggs, High resolution XPS of organic polymers, in: The Scienta ESCA 300 Database, John Wiley and Sons, Chichester, 1992.

40. V. Crist, Handbook of Monochromatic XPS Spectra (Elements and Native Oxides), XPS International LLC, Mountain View, CA, USA, 2004.

41. E. Atanassova, D. Spassov, X - ray photoelectron spectroscopy of thermal thin Ta2O5 films on Si, Applied Surface Science 135 (1998) 71.

42. A. Muto, F. Yano, Y. Sugawara, S. Iijima, The study of ultrathin tantalum oxide films before and after annealing with X-ray photoelectron spectroscopy, Journal of Applied Physics 33 (1994) 2699.

43. J.A. Rotole, P.M.A. Sherwood, Gamma-Alumina (alpha-Al2O3) by XPS, Surface Science Spectra 5 (1998) 18.

44. J. Theo Kloprogge, L.V. Duong, B.J. Wood, R.L. Frost, XPS study of the major minerals in bauxite, Journal of Colloid and Interface Science 296 (2006) 572.

45. V. Maurice, W.P. Yang, P. Marcus, XPS and STM investigation of the passive film formed on Cr(110) single crystal surfaces, Journal of the Electrochemical Society 141 (1994) 3016.

46. J.T. Li, V. Maurice, J. Swiatowska-Mrowiecka, A. Seyeux, S. Zanna, L. Klein, S.G. Sun, P. Marcus, XPS, Time-of-Flight-SIMS and Polarization Modulation IRRAS study of Cr2O3 thin film materials as anode for lithium ion battery, Electrochimica Acta 54 (2009) 3700.

47. D.F. Mitchell, G.I. Sproule, M.J. Graham, Sputter reduction of oxides by ion bombardment during Auger depth profile analysis, Surface and Interface Analysis 15 (1990) 487.

48.   M. Descostes, F. Mercier, N. Thromat, C. Beaucaire, M. Gautier-Soyer, Use of XPS in the determination of chemical environment and oxidation state of iron and sulfur samples: constitution of a data basis in binding energies for Fe and S reference compounds and applications to the evidence of surface species of an oxidized pyrite in a carbonate medium, Applied Surface Science 165 (2000) 288.

49.   A.P. Grosvenor, B.A. Kobe, N.S. McIntyre, Studies of the oxidation of iron by water vapour using X-ray photoelectron spectroscopy and QUASES, Surface Science 572 (2004) 217.

50.   A.P. Grosvenor, B.A. Kobe, M.C. Biesinger, N.S. McIntyre, Investigation of multiplet splitting of Fe 2p XPS spectra and bonding in iron compounds, Surface and Interface Analysis 36 (2004) 1564.

51.   D. Briggs, M.P. Seah, Practical Surface Anlaysis, Vol. 1, 2nd edition, John Wiley & Sons, Chichester, 1993.

52.   V. Maurice, S. Cadot, P. Marcus, XPS, LEED and STM study of thin oxide films formed on Cr(110), Surface Science 458 (2000) 195.

53.   V. Maurice, S. Cadot, P. Marcus, Hydroxylation of ultra-thin films of Cr2O3(0001) formed on Cr(110), Surface Science 471 (2001) 43.

54.   D.A. López, W.H. Schreiner, S.R. de Sánchez, S.N. Simison, The influence of carbon steel microstructure on corrosion layers: an XPS and SEM characterization, Applied Surface Science 207 (2003) 69.

55.   J. Walter, W. Boonchuduang, S. Hara, XPS study on pristine and intercalated tantalum carbosulfide, Journal of Alloys and Compounds 305 (2000) 259.

56.   S. Joiret, M. Keddam, X.R. Nóvoa, M.C. Pérez, C. Rangel, H. Takenouti, Use of EIS, ring-disk electrode, EQCM and Raman spectroscopy to study the film of oxides formed on iron in 1 M NaOH, Cement and Concrete Composites 24 (2002) 7.

57.   P.M. Natishan, E. McCafferty, P.R. Puckett, S. Michel, Ion beam assisted deposited tantalum oxide coatings on aluminum,

Corrosion Science 38 (1996) 1043.

58. P. Panjan, M. Cekada, M. Panjan, D. Kek-Merl, Growth defects in PVD hard coatings, Vacuum 84 (2010) 209.

59. CRC Handbook of Chemistry and Physics, 55th edition, CRC Press, 1974–1975.

60. E. Barsoukov, J.R. Macdonald (Eds.), Impedance Spectroscopy Theory, Experiment, and Applications, 2nd edition, John Wiley & Sons, Chichester, 2005.

61. G.J. Brug, A.L.G. van den Eeden, M. Sluyters-Rehbach, J.H. Sluyters, The analysis of electrode impedances complicated by the presence of a constant phase element, Journal of Electroanalytical Chemistry and Interfacial Electrochemistry 176 (1984) 275.

62. V.M-W. Huang, V. Vivier, M.E. Orazem, N. Pébère, B. Tribollet, The apparent constant-phase-element behavior of a disk electrode with faradaic reactions: a global and local impedance analysis, Journal of the Electrochemical Society 154 (2007) C99.

63. C.M.Abreu,M. Izquierdo,M.Keddam,X.R. Nóvoa, H. Takenouti, Electrochemical behaviour of zinc-rich epoxy paints in 3% NaCl solution, Electrochimica Acta 41 (1996) 2405.

64. I. Epelboin, M. Keddam, J.C. Lestrade, Faradaic impedances and intermediates in electrochemical reactions, Faraday Discussions of the Chemical Society 56 (1973) 264.

65. J.W. Lenderink, M.V.D. Linden, J.H.W. de Wit, Corrosion of aluminium in acidic and neutral solutions, Electrochimica Acta 38 (14) (1993) 1989.

66. P. Li, T.C. Tan, J.Y. Lee, Impedance spectra of the anodic dissolution of mild steel in sulfuric acid, Corrosion Science 38 (1996) 1935.

67. G.A. Zhang, Y.F. Cheng, On the fundamentals of electrochemical corrosion of X65 steel in CO2-containing formation water in the presence of acetic acid in petroleum production, Corrosion Science 51 (2009) 87.

68. G.A. Zhang, Y.F. Cheng, Corrosion of X65 steel in CO2-saturated oilfield formation water in the absence and presence of acetic acid, Corrosion Science 51 (2009) 1589.

69. F. Farelas, M. Galicia, B. Brown, S. Nesic, H. Castaneda, Evolution of dissolution processes at the interface of carbon steel corroding in a CO2 environment studied by EIS, Corrosion Science 52 (2010) 509.

70. M.A. Jingling, W. Jiuba, L.I. Gengxin, X.V. Chunhua, The corrosion behaviour of Al–Zn–In–Mg–Ti alloy in NaCl solution, Corrosion Science 52 (2010) 534.

71. F. Wenger, S., Cheriet, B. Talhi, J. Galland, Electrochemical impedance of pits. Influence of the pit morphology, Corrosion Science 39 (7) (1997) 1239.

72. K. Kukli, M. Ritala, M. Leskelä, Atomic layer epitaxy growth of tantalum oxide thin-films from Ta(OC2H5)(5) and H2O, Journal of the Electrochemical Society 142 (1995) 1670.

73. G.E. Cavigliasso, M.J. Esplandiu, V.A. Macagno, Influence of the forming electrolyte on the electrical properties of tantalum and niobium oxide films: an EIS comparative study, Journal of Applied Electrochemistry 28 (1998) 1213.

74. E.M.A. Martini, I.L. Muller, Characterization of the film formed on iron in borate solution by electrochemical impedance spectroscopy, Corrosion Science 42 (2000) 443.

75. Y.-Y. Chang, D.-Y. Wang, Corrosion behavior of electroless nickel-coated AISI 304 stainless steel enhancedby titaniumionimplantation, Surface andCoatings Technology 200 (2005) 2187.

Chapter 4

Anodic Dissolution of API X70 Pipeline Steel in Arabian Gulf Seawater after Different Exposure Intervals

El-Sayed M. Sherif[1, 2] and Abdulhakim A. Almajid[1]

[1]Department of Mechanical Engineering, College of Engineering, King Saud University, Riyadh 11421, Saudi Arabia

[2]Electrochemistry and Corrosion Laboratory, Department of Physical Chemistry, National Research Centre (NRC), Dokki, Giza 12622, Egypt

ABSTRACT

The anodic dissolution of API X70 pipeline steel in Arabian Gulf seawater (AGSW) was investigated using open-circuit potential (OCP), electrochemical impedance spectroscopy (EIS), cyclic potentiodynamic polarization (CPP), and current-time measurements. The electrochemical experiments revealed that the X70 pipeline steel suffers both general and pitting corrosion in the AGSW solution. It was found that the general corrosion decreases as a result of decreasing the corrosion current density (J_{corr}), corrosion rate (R_{corr}) and absolute currents as well as the increase of polarization resistance of X70 with increasing the exposure time. On the other hand, the pitting corrosion was found to increase with increasing the immersion time. This was confirmed by the increase of current with time and by the SEM images that were obtained on the steel surface after 20 h immersion before applying an amount of 0–.35 V versus Ag/AgCl for 1 h.

INTRODUCTION

API X70 pipeline steel is characterized by its good combination of strength and toughness, good weldability, low crack sensitivity coefficient, and low ductile to brittle transition temperature [1]. Therefore, several studies [1–5] have been conducted on the corrosion and corrosion protection of X70 pipeline steel in different aggressive media. Alizadeh and Bordbar have [1] reported the influence of microstructure on the protective properties of the corrosion product layer generated on the welded API X70 steel in sodium chloride solutions and found that the corrosion resistance for the steel was increased after heat treatment due to formation of fine and compact corrosion product layer with fewer defects. Bordbar and Alizadeh [2] also investigated the effects of microstructure alteration on corrosion behavior of welded joint in API X70 pipeline steel. Li et al. [3] studied the anodic dissolution of the X70 steel in H_3PO_4 solution with the halide ion perturbation at the interface and reported that the way how the halide ions affect the anodic dissolution is related to the types of the ions and the property of the film formed on the surface of the electrode. The corrosion of welded X70 pipeline steel in near-neutral pH solution was also characterized by Zhang and Cheng [4] using the microelectrochemical technique and confirmed that the resistance of corrosion product layer of the steel decreases with hydrogen charging and heat-affected zone has the largest dissolution current upon hydrogen charging. Furthermore, Sirong et al. [5] have invented a new method for preparing bionic multiscale superhydrophobic functional surface on X70 pipeline steel.

It is generally believed that general and pitting corrosion of passivated pipeline steel often occur when this steel is in contact with aggressive ions such as the chloride ions. In this regard, the corrosion of steel in various media such as carbonate/bicarbonate solution [6–9] and acidic solutions [3, 10, 11] has been reported. Some other investigators have focused their studies on the corrosion of steels under disbanded coating [12–14] and hydrogen damage [15]. Although there are a lot of investigations on the corrosion of API X70 pipeline steel, the corrosion of this steel in Arabian Gulf seawater (AGSW) has not yet been reported to the best of our knowledge. The aim of this work was to investigate the corrosion of API X70 pipeline steel after its immersion in the freely aerated AGSW for different exposure periods, namely, 1 h and 20 h, using varied electrochemical techniques such as potential-time, electrochemical impedance spectroscopy, cyclic polarization, and potentiostatic current-time measurements along with scanning electron microscopy investigations. A particular attention was paid to the effect of immersion time on the pitting corrosion of the API X70 pipeline steel.

MATERIALS AND METHODS

Materials and Chemicals

API X70 pipeline steel with a rectangular shape and dimension of 1 cm length and 1 cm width was employed in this study. The main chemical compositions for this steel were 0.04% C, 1.70% Mn, 0.035% S_{max}, 0.035% P_{max}, and 0.55% Si_{max}; all these elements were in mass percent. Arabian Gulf seawater was brought from the Arabian Gulf at Dammam, Saudi Arabia.

Electrochemical Corrosion Measurements

The electrochemical measurements were collected using a conventional electrochemical cell with a three-electrode configuration. The API X70 pipeline steel rod, a platinum foil, and an Ag/AgCl electrode (in saturated KCl solution) were the working, counter, and reference electrodes, respectively. Before measurements, the surface of the working electrode was ground successively with metallographic emery paper of increasing fineness up to 1200 grit; it was then cleaned using doubly distilled water, degreased with acetone, washed using doubly distilled water again, and finally dried with dry air.

The electrochemical measurements were carried out using an Autolab Potentiostat (PGSTAT20 computer controlled) operated by the general purpose electrochemical software (GPES) version 4.9. The electrochemical impedance spectroscopy (EIS) tests were performed at open-circuit potential (OCP) over a frequency range of 100 kHz to 100 mHz, with an ac wave of ±5 mV peak-to-peak overlaid on a dc bias potential, and the impedance data were collected using Powersine software at a rate of 10 points per decade change in frequency. The cyclic potentiodynamic polarization curves were obtained between −1.2 and 0.25 potential rang at a scan rate of 1 mV/s versus Ag/AgCl. Potentiostatic current-time experiments were carried out by stepping the potential of the steel samples at −0.350 V versus Ag/AgCl. All experiments were carried out using a fresh electrode surface at room temperature in a cell that contains 200 mL of the test solutions.

Scanning Electron Microscope (SEM) Investigations

The SEM images were obtained by using a JEOL model JSM-6610LV

(Japanese made) scanning electron microscope with an energy dispersive X-ray analyzer attached for acquiring the EDX analysis.

RESULTS AND DISCUSSION

Open-Circuit Potential (OCP) Measurements

Figure 1 shows the variation of the potential of API X70 pipeline steel in AGSW against time for circa 20 h. It is seen that the potential of the steel in AGSW increased towards the more negative values from the first moment of electrode immersion, which resulted from the dissolution of an air oxide film which was formed on the steel surface before its immersion. The potential further decreased in the more negative direction due to the continuous dissolution of the steel surface under the aggressiveness action of the anions present in the AGSW. After about 4 h, the steel potential abruptly shifted in the less negative direction indicating that the steel surface developed an oxide film and/or corrosion product layer that could provide certain protection and decreased the corrosive attack of the AGSW on the surface. The potential then continued shifting in the less negative direction with time till the end of the run. This was due to the oxide film and/or corrosion product thickening with time, which provided more protection for the steel surface and led to the increase of potential towards the less negative values.

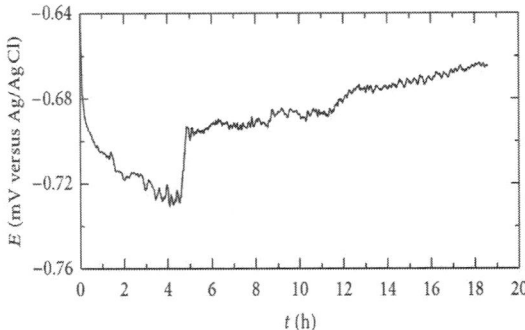

Figure 1: Change of the open-circuit potential versus time for the API X70 pipeline steel in Arabian Gulf seawater.

Electrochemical Impedance Spectroscopy (EIS) Measurements

Figure 2 shows the EIS Nyquist plots obtained for the API X70 pipeline steel after its immersion for (a) 1 hour and (b) 20 h in Arabian Gulf seawater. The data obtained from Figure 2 were best fitted to the equivalent circuit model depicted in Figure 3. The values of the elements of the equivalent circuit shown in Figure 3 are listed in Table 1, where R_s represents the solution resistance, Q the constant phase elements (CPEs), and R_p the polarization resistance [16–19].

Table 1: EIS parameters obtained by fitting the Nyquist plots shown in Figure 2 with the equivalent circuit shown in Figure 3 for X70 steel after 1 h and 20 h immersion in Arabian Gulf seawater

Medium	Parameter		
	R_s (Ω cm²)	Q(CPEs)		R_p (Ω cm²)
		Y_0/F cm⁻²	n	
AGSW (1 h)	12.24	0.001433	0.78	839
AGSW (20 h)	13.12	0.0001047	0.80	1025

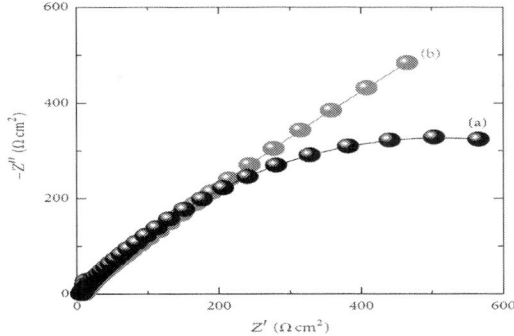

Figure 2: EIS Nyquist plots obtained for the API X70 pipeline steel after its immersion for (a) 1 h and (b) 20 h in Arabian Gulf seawater.

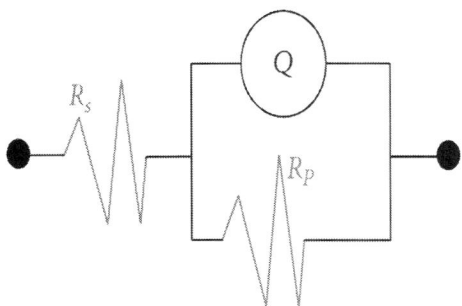

Figure 3: The equivalent circuit model used to fit the experimental data presented by Nyquist plots shown in Figure 2; symbols of the circuit are defined in the text.

It is clearly seen from Figure 2 that there is only one semicircle whether the immersion time was 1 h or 20 h. The diameter of the semicircle is higher for the steel that was immersed for 20 h, curve (b), than that obtained after 1 h. This indicates that increasing the immersion time decreases the aggressiveness action of the AGSW towards the steel, which is due to the fact that the increased immersion time allows the surface to develop a corrosion product layer and/or an oxide film that gets thicker with time and provides some protection to the surface against corrosion. This was also confirmed by the data recorded in Table 1, where the R_s and R_p values are higher for the steel that was immersed for 20 h in AGSW before measurements. The constant phase elements (CPEs, Q) with their n values are around 0.8; CPEs thus represent double layer capacitors because their n values are close to unity. The value of Q decreased with increasing time, which proves that the mass transport from surface is limited, particularly after 20 h of the steel immersion in the AGSW [20–22]. According to Arzola-Peralta et al. [23], who studied the corrosion of X70 pipeline steel at similar conditions, at high frequencies a transfer process takes place. This emphasizes the fact that increasing the immersion time before measurements decreases the corrosion of the steel, which can result from the formation of an oxide film and/or corrosion products that could partially protect the surface from being attacked by the chloride ions present in solutions under investigations.

The data obtained from Nyquist plots were also confirmed by the Bode impedance and Bode phase angle plots that are shown in Figures 4 and 5, respectively, for the API X70 pipeline steel electrode after its immersion in AGSW for (a) 1 h and (b) 20 h. Figure 4 depicts that the impedance of the interface for the X70 increases with increasing the immersion time from 1 h to 20 h, particularly at lower frequency values. According to Mansfeld et al. [22], the high impedance values at the low frequency region confirms the passivation of the surface. This indicates that the increase of the immersion time from 1 h to 20 h increases the corrosion resistance of the steel in AGSW through increasing the passivity of its surface. Figure 5 also showed that the increase of the immersion time from 1 h to 20 h before measurements increased the maximum degree of the phase angle, which gives further confirmation on the increased surface passivity with increased exposure intervals.

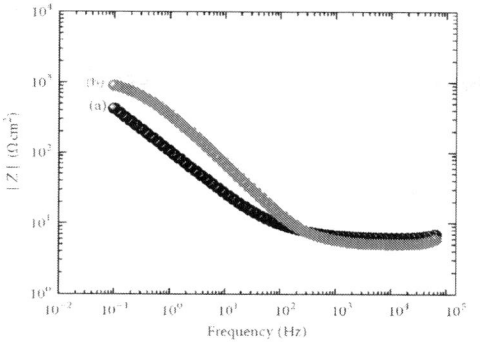

Figure 4: Bode impedance plots obtained for the API X70 pipeline steel after its immersion for (a) 1 h and (b) 20 h in Arabian Gulf seawater.

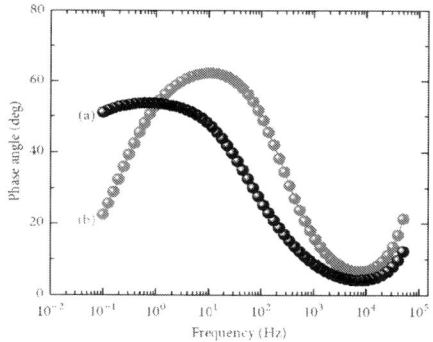

Figure 5: Bode phase angle plots obtained for the API X70 pipeline steel after its immersion for (a) 1 h and (b) 20 h in Arabian Gulf seawater.

Cyclic Potentiodynamic Polarization (CPP) Data

The CPP curves obtained for API X70 pipeline steel after its immersion for (a) 1 h and (b) 20 h in the Arabian Gulf seawater are shown, respectively, in Figure 6. These curves were collected in order to understand the change of current with potential and to report the corrosion parameters for the API X70 steel after varied exposure intervals in AGSW. The values of the corrosion parameters such as cathodic (β_c) and anodic (β_a) Tafel slope, corrosion potential (E_{Corr}), corrosion current density (j_{Corr}), pitting potential (E_{Pit}), pitting current density (j_{Pit}), protection potential (E_{Prot}), polarization resistance (R_p), and corrosion rate (R_{Corr}) obtained for the API X70 steel electrodes from the CPP curves shown in Figure 6 are listed in Table 2. The values of all these parameters were obtained as reported in the previous studies [16–19].

Table 2: Parameters obtained from cyclic potentiodynamic polarization curves shown in Figure 6 for X70 steel after 1 h and 20 h immersion in Arabian Gulf seawater

Medium	Parameter							
	ECorr(mV)	JCorr (μ A cm−2)	β_C (mVdec−1)	β_a (mVdec−1)	EPit (mV)	EProt (mV)	RP(kΩcm2)	RCorr (mmy−1)
AGSW (1 h)	−850	16	95	450	−450	−600	2.13	0.167
AGSW (20h)	−860	13	100	200	−375	−575	2.22	0.151

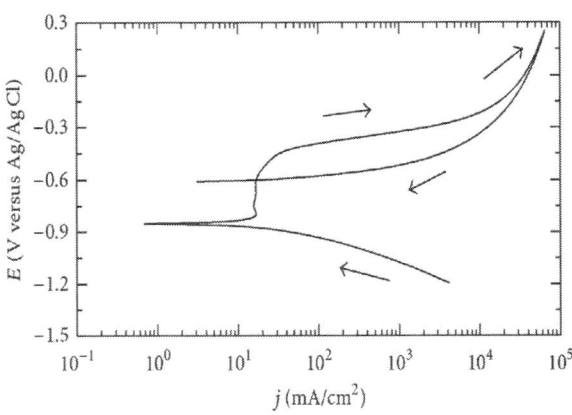

(a)

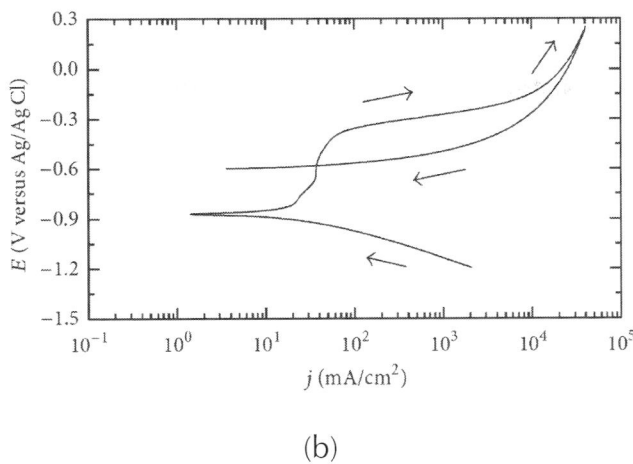

(b)

Figure 6: Cyclic potentiodynamic polarization curves obtained for API X70 pipeline steel after its immersion for (a) 1 h and (b) 20 h in the Arabian Gulf seawater.

It has been reported [24–28] that the cathodic reaction of metals and alloys in an open to air near neutral solution is the oxygen reduction. On the other hand, the anodic reaction takes place via the dissolution of iron from Fe(0) to Fe(II). According to Alizadeh and Bordbar [1], the cathodic and anodic reactions can be represented, respectively, as follows:

$$O_2 + 2H_2O + 4e \rightleftarrows 4OH^- \tag{1}$$

$$Fe \rightleftarrows Fe^{2+} + 2e \tag{2}$$

In the chloride containing carbonate/bicarbonate, which is the case of the natural seawater, AGSW, the $FeCO_3$ can be formed and deposited on the steel surface as follows [1]:

$$Fe^{2+} + CO_3{}^{2-} \rightleftarrows FeCO_3 \tag{3}$$

$$Fe + HCO_3{}^- + e \rightleftarrows FeCO_3 + H \tag{4}$$

Fu and Cheng [29] have also reported that the formation and the deposition of $FeCO_3$ provide an inhibition for the steel surface from further dissolution.

It is clearly seen from Figure 6 that the X70 pipeline steel shows an active-passive anodic behavior, where the steel forms a passive region in the potential range between −800 and −400 mV for the steel in the AGSW solution; the appearance of such region is due to the formation of an oxide film and/or corrosion products. At these conditions, the corrosion products might be iron carbonate and chloride compounds such as $FeCO_3$, $FeCl_2$, and $FeOCl$ and the oxides can be Fe_3O_4 and Fe_2O_3 [1]. The formation of such oxides can be represented as follows [30]:

$$4FeCO_3 + O_2 + 4H_2O \rightleftarrows 2Fe_2O_3 + 4HCO_3^- + 4H^+$$

(5)

$$6FeCO_3 + O_2 + 6H_2O \rightleftarrows 2Fe_3O_4 + 6HCO_3^- + 6H^+$$

(6)

Increasing the applied potential in the less negative direction leads to increasing the output current due to the dissolution of the formed passive film and the occurrence of pitting corrosion. Reversing the potential in the backward scan resulted in an increase in the output current and the appearance of a hysteresis loop, which indicates the occurrence of pitting corrosion for the API X70 steel in the AGSW test medium. Table 2 also indicated that the j_{Corr}, anodic current, and R_{Corr} recorded higher values, while the (R_p) is lower for the steel in AGSW after 1 h immersion. Increasing the immersion time from 1 h to 20 h decreased the values of j_{Corr} and R_{Corr} and increased the values of (R_p). This also agrees with the impedance data that the corrosion of API X70 steel decreases with increasing the immersion time before measurements.

Potentiostatic Current-Time and Scanning Electron Microscopy Investigations

The variation of the potentiostatic current versus time at −0.350 mV versus Ag/AgCl after (a) 1 h and (b) 20 h in AGSW is shown in Figure 7. These experiments were carried out to report the anodic dissolution of the API X70 steel at an active anodic potential and to see whether pitting corrosion occurs after varied exposure periods in the test solution. The current of the steel after 1 h immersion in AGSW, curve (a), recorded very high value that increased in the first few seconds before decreasing again till the first 400 s. The increase of current might have resulted from the dissolution of an oxide film which was formed on the surface due to the contact with the solution. On the other hand, the decrease in the current values was either due to the formation of an oxide and/or corrosion products film that decreases the effect of the aggressiveness action of the test solution. The current then shows almost a stable change with increasing time up to the end of the run.

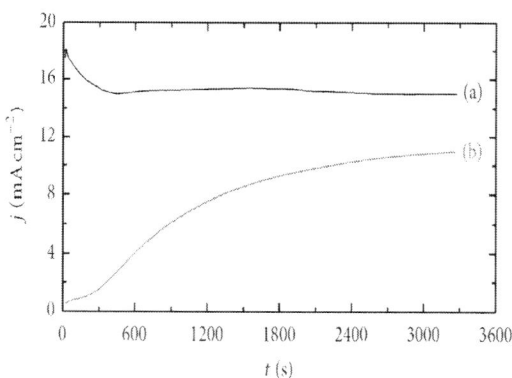

Figure 7: Potentiostatic current-time curves obtained at −0.35 V versus Ag/AgCl for the API X70 pipeline steel electrode after (a) 1 hour and (b) 20 h immersions in the Arabian Gulf seawater.

Increasing the immersion time to 20 h before measurements could decrease the initial currents for steel in the AGSW solution to its minimum due to the formation of corrosion products and/or passive film on the electrode surface during its immersion before applying the constant potential. The current then gradually increases with increasing the time of the experiment as a result of the dissolution of the formed passive film and the occurrence of pitting corrosion. This can be also explained by the dissolution of iron as represented by (2) and then the formation of $FeCl_2$ and $FeCl_3$ on the surface of the steel according to the following reactions [31, 32]:

$$Fe_{(s)} + 2Cl^-_{(aq)} \rightleftharpoons FeCl_{2(s)} + 2e^- \tag{7}$$

$$FeCl_{2(s)} + Cl^-_{(aq)} \rightleftharpoons FeCl_{3(s)} + e^- \tag{8}$$

Due to the applied potential and concentration gradients, the $FeCl_2$ and $FeCl_3$ species at the interface diffuse through the porous film and the diffusion boundary layer and are then carried away to the bulk solution leading to a continuous dissolution of the alloy and the occurrence of pitting corrosion as indicated by the continuous increase of the current with time.

This was confirmed by the SEM micrographs obtained for the API X70 pipeline steel surface after performing the current-time experiment shown in Figure 7 (curve (b)). Figure 8 shows the SEM micrographs obtained for the API X70 pipeline steel surface after its immersion for 20 h then applying −0.35 mV for 1 h, which shows an increased magnification for the steel surface. Figure 8(a) depicts that the majority of the steel surface has few pits and some of these pits were propagated and could lead to the continuous increase of the current with time which we have seen on curve (b) of Figure 7. Figures 8(b) and 8(c) show clear images of the propagated pits and prove that the increase of current with time at −0.35 V was due to the occurrence of pitting corrosion. The data obtained by potentiostatic current-time thus confirm the data obtained by EIS and polarization measurements that increasing the immersion time of the API X70 pipeline steel decreases the uniform corrosion and increases the pitting one.

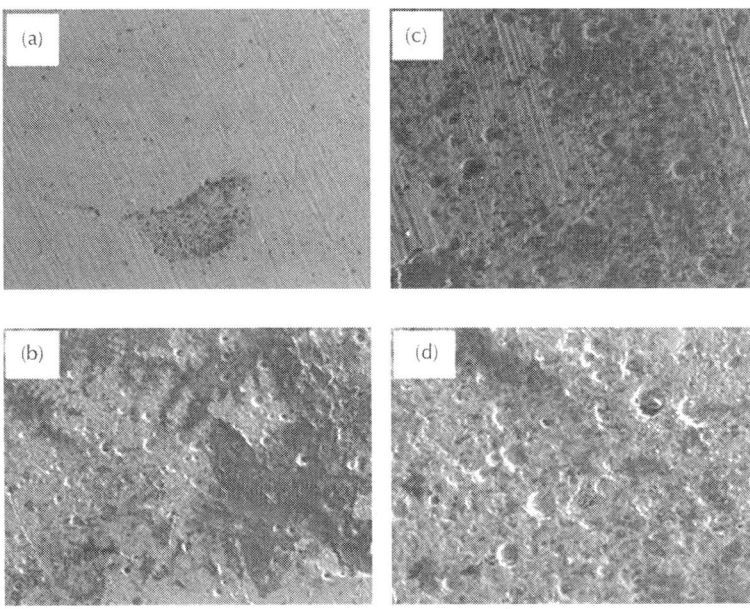

Figure 8: SEM micrographs obtained on the API X70 pipeline steel electrode after its immersion for 20 h in the Arabian Gulf seawater then stepping the potential to −0.35 V versus Ag/AgCl for 1 h.

CONCLUSIONS

The corrosion of API X70 pipeline steel after 1 h and 20 h immersion in Arabian Gulf seawater (AGSW) has been reported using variety of electrochemical measurements. Potential-time experiments indicated that the potential of the steel shifts to the more negative values in the first few minutes of the immersion and then forms a layer of oxides and/or corrosion products that partially protect the steel surface and shift its potential towards the less negative values with time. Electrochemical impedance spectroscopy data revealed that X70 steel shows one semicircle in AGSW solution and increasing the immersion time to 20 h increases the surface and polarizations resistances. Cyclic polarization technique confirmed that the steel shows lower corrosion current and corrosion rate and higher polarization resistance in AGSW with the increase of the immersion time from 1 h to 20 h, which indicates that the general corrosion decreases and the pitting corrosion increases. Potentiostatic current-time experiments at −0.35 mV versus Ag/AgCl indicated that the X70 steel confirmed the data obtained by EIS and polarization measurements that the current recorded higher values with 1 h exposure period. On the other hand, increasing the time to 20 h decreased the absolute currents of the steel, where the initial currents were very low and increased with increasing the time of the applied potential. All measurements thus were consistent with each other and proved that the increase of the immersion time of the API X70 steel in the AGSW solution before measurements decreases the uniform corrosion through decreasing the absolute current values, while it increases the pitting corrosion by increasing the current with time.

ACKNOWLEDGMENTS

The authors extend their appreciation to the Deanship of Scientific Research at KSU for funding the work through the research group Project no. RGP-VPP-160.

REFERENCES

1.   M. Alizadeh and S. Bordbar, "The influence of microstructure on the protective properties of the corrosion product layer generated on the welded API X70 steel in chloride solution," Corrosion Science, vol. 70, p. 170, 2013

2.   S. Bordbar and M. Alizadeh, "Effects of microstructure alteration on corrosion behavior of welded joint in API X70 pipeline steel," Materials and Design, vol. 45, p. 597, 2013

3.   L. Li, C. Wang, and H. Lu, "Anodic dissolution of the X70 steel in H_3PO_4 solution with the halide-ion perturbation at the

interface," Electrochimica Acta, vol. 104, p. 295, 2013

4. G. A. Zhang and Y. F. Cheng, "Micro-electrochemical characterization of corrosion of welded X70 pipeline steel in near-neutral pH solution," Corrosion Science, vol. 51, no. 8, pp. 1714–1724, 2009

5. Y. Sirong, W. Xiaolong, W. Wei, Y. Qiang, X. Jun, and X. Wei, "A new method for preparing bionic multi scale superhydrophobic functional surface on X70 pipeline steel," Applied Surface Science, vol. 271, p. 149, 2013

6. M. Torabi Merajin, S. Sharifnia, S. N. Hosseini, and N. Yazdanpour, "Photocatalytic conversion of greenhouse gases (CO_2 and CH_4) to high value products using TiO_2 nanoparticles supported on stainless steel webnet," Journal of the Taiwan Institute of Chemical Engineers, vol. 44, p. 239, 2013

7. A. A. Oskuie, T. Shahrabi, A. Shahriari, and E. Saebnoori, "Electrochemical impedance spectroscopy analysis of X70 pipeline steel stress corrosion cracking in high pH carbonate solution," Corrosion Science, vol. 61, p. 111, 2012

8. S.-Y. Zhao, S.-H. Chen, H.-Y. Ma, D.-G. Li, and F.-J. Kong, "Current oscillations during electrodissolution of iron in perchloric acid solutions," Journal of Applied Electrochemistry, vol. 32, no. 2, pp. 231–235, 2002

9. Z. Y. Liu, X. G. Li, C. W. Du, and Y. F. Cheng, "Local additional potential model for effect of strain rate on SCC of pipeline steel in an acidic soil solution," Corrosion Science, vol. 51, no. 12, pp. 2863–2871, 2009

10. S. Alarmal Mangai and S. Ravi, "Comparative corrosion inhibition effect of imidazole compounds and of Trichodesma indicum (Linn) R. Br. on C38 steel in 1 M HCl medium," Journal of Chemistry, vol. 2013, Article ID 527286, 4 pages, 2013

11. D. Sazou and M. Pagitsas, "Nitrate ion effect on the passive film breakdown and current oscillations at iron surfaces polarized in chloride-containing sulfuric acid solutions," Electrochimica Acta, vol. 47, no. 10, pp. 1567–1578, 2002

12. A. Q. Fu, X. Tang, and Y. F. Cheng, "Characterization of corrosion of X70 pipeline steel in thin electrolyte layer under disbonded coating by scanning Kelvin probe," Corrosion Science, vol. 51, no. 1, pp. 186–190, 2009

13. M. Yan, J. Wang, E. Han, and W. Ke, "Local environment under simulated disbonded coating on steel pipelines in soil solution," Corrosion Science, vol. 50, no. 5, pp. 1331–1339, 2008

14. J. Min, J. H. Park, H.-K. Sohn, and J. M. Park, "Synergistic effect of potassium metal siliconate on silicate conversion coating for corrosion protection of galvanized steel," Journal of Industrial and Engineering Chemistry, vol. 18, no. 2, pp. 655–660, 2012

15. H. C. Genuino, N. N. Opembe, E. C. Njagi, S. McClain, and S. L. Suib, "A review of hydrofluoric acid and its use in the car wash industry," Journal of Industrial and Engineering Chemistry, vol. 18, p. 1529, 2012

16. E. M. Sherif and S.-M. Park, "Inhibition of copper corrosion in 3.0% NaCl solution by N-phenyl-1,4-phenylenediamine," Journal of the Electrochemical Society, vol. 152, no. 10, pp. B428–B433, 2005

17. E. M. Sherif and S.-M. Park, "Effects of 2-amino-5-ethylthio-1,3,4-thiadiazole on copper corrosion as a corrosion inhibitor in aerated acidic pickling solutions," Electrochimica Acta, vol. 51, no. 28, pp. 6556–6562, 2006

18. E. M. Sherif and S.-M. Park, "Inhibition of copper corrosion in acidic pickling solutions by N-phenyl-1,4-phenylenediamine," Electrochimica Acta, vol. 51, no. 22, pp. 4665–4673, 2006

19. E. M. Sherif and S.-M. Park, "2-Amino-5-ethyl-1,3,4-thiadiazole as a corrosion inhibitor for copper in 3.0% NaCl solutions," Corrosion Science, vol. 48, no. 12, pp. 4065–4079, 2006

20. J. R. Macdonald, Impedance Spectroscopy, John Wiley & Sons, New York, NY, USA, 1987.

21. P. C. Okafor, C. B. Liu, X. Liu et al., "Corrosion inhibition and

adsorption behavior of imidazoline salt on N80 carbon steel in CO_2-saturated solutions and its synergism with thiourea," Journal of Solid State Electrochemistry, vol. 14, no. 8, pp. 1367–1376, 2010

22. F. Mansfeld, S. Lin, K. Kim, and H. Shih, "Pitting and surface modification of SIC/Al," Corrosion Science, vol. 27, no. 9, pp. 997–1000, 1987

23. S. Arzola-Peralta, J. Genesca-Llongueras, J. Mendoza-Flores, and R. Duran-Romero, "Corrosion behavior of X70 pipeline steel in H_2S containing solutions," in Proceedings of the Corrosion, NACE International, New Orleans, La, USA, April 2004

24. E. J. Calvo and N. Mozhzhukhina, "A rotating ring disk electrode study of the oxygen reduction reaction in lithium containing non aqueous electrolyte," Electrochemistry Communication, vol. 31, p. 56, 2013.

25. V. P. Shinde and P. P. Patil, "A study on the electrochemical polymerization, characterization, and corrosion protection of o-toluidine on steel," Journal of Solid State Electrochemistry, vol. 19, p. 29, 2013.

26. E.-S. M. Sherif, R. M. Erasmus, and J. D. Comins, "Corrosion of copper in aerated synthetic sea water solutions and its inhibition by 3-amino-1,2,4-triazole," Journal of Colloid and Interface Science, vol. 309, no. 2, pp. 470–477, 2007

27. M. El-Sayed Sherif, A. A. Almajid, K. A. Khalil, H. Junaedi, and F. H. Latief, "Electrochemical Studies on the Corrosion Behavior of API X65 Pipeline Steel in Chloride Solutions," International Journal of Electrochemical Science, vol. 8, p. 9360, 2013

28. E.-S. M. Sherif and A. H. Ahmed, "Synthesizing new hydrazone derivatives and studying their effects on the inhibition of copper corrosion in sodium chloride solutions," Synthesis and Reactivity in Inorganic, Metal-Organic and Nano-Metal Chemistry, vol. 40, no. 6, pp. 365–372, 2010

29. Q. A. Fu and Y. F. Cheng, "Electrochemical polarization

behavior of X70 steel in thin carbonate/bicarbonate solution layers trapped under a disbonded coating and its implication on pipeline SCC," Corrosion Science, vol. 52, p. 2511, 2010

30. J. K. Heuer and J. F. Stubbins, "An XPS characterization of $FeCO_3$ films from CO_2 corrosion," Corrosion Science, vol. 41, no. 7, pp. 1231–1243, 1999

31. C. A. Melendres, N. Camillone III, and T. Tipton, "Laser raman spectroelectrochemical studies of anodic corrosion and film formation on iron in phosphate solutions," Electrochimica Acta, vol. 34, no. 2, pp. 281–286, 1989.

32. J. L. Yao, B. Ren, Z. F. Huang, P. G. Cao, R. A. Gu, and Z.-Q. Tian, "Extending surface Raman spectroscopy to transition metals for practical applications IV. A study on corrosion inhibition of benzotriazole on bare Fe electrodes," Electrochimica Acta, vol. 48, no. 9, pp. 1263–1271, 2003

A study on Cathodic Protection against Crevice Corrosion in Dilute NaCl Solutions

Zhengfeng Li[a, b], Fuxing Gan[a, b], and Xuhui Mao[a, b]

[a]College of Chemistry and Environmental Science, Wuhan University, Wuhan 430072, People's Republic of China
[b]State Key Laboratory for Corrosion and Protection of Metals, Shenyang 110015, People's Republic of China

ABSTRACT

Potential and current distributions in a cathodically protected crevice between a simulated coating and segmented mild steel electrodes were measured in dilute NaCl solutions. The distributions became more uniform with time due to an increase in solution conductivity and depletion of dissolved oxygen in the crevice. Generally, a negative shift of control potential and an increase in initial solution conductivity and crevice thickness resulted in a higher polarization level on the steel. However, if the control potential is too negative, the polarization level may be lower than that under a suitable control potential because of hydrogen evolution. On the basis of these results, a mechanism of cathodic protection against crevice corrosion in high-resistivity environments was proposed.

INTRODUCTION

Buried or submerged steel structures, especially steel pipelines, are commonly prevented from external corrosion by a combined application of coatings and cathodic protection (CP). Dielectric coatings can isolate the external surface of steel structures from corrosive environments. CP can complete the protection system by preventing corrosion of steel surface at holidays, namely pinholes and ruptures, which would develop on the coatings through aging process. However, crevices are apt to form beneath the coatings in the surrounding areas of holidays. Water then flows through the holiday opening into the crevices, and crevice corrosion may occur [1], [2] and [3]. Thus, there is a question to be answered: can CP prevent the crevice corrosion?

Potential distributions in cathodically protected crevices were measured in low-resistivity environments by some research groups [3], [4], [5] and [6]. Their results indicate it is possible to achieve CP on the whole steel surface in the crevice containing a low-resistivity electrolyte. However, very limited literature information is available about potential distributions, especially current distributions in a cathodically protected crevice in high-resistivity environments [7] and [8]. Gan et al. [9] and Jack et al. [10] measured respectively potential distributions within a cathodically protected crevice between a simulated disbonded coating and pipeline steel in dilute chloride solutions with high resistivity. They found that it was hard to polarize the steel surface in the crevice to a potential more negative than −0.85 V vs. a saturated copper–cupric sulfate electrode (CSE) or −0.77 V vs. a saturated calomel electrode (SCE). Similar measurements were performed by Brousseau and Qian [11] in aqueous solution with a high resistivity of 5070 Ω cm. The results showed that a more negative control potential at holiday prompted higher cathodic polarization level along the entire crevice. However, Fessler et al. [12] concluded that control potentials more negative than that required to produce hydrogen bubbles would cause difficulty in controlling the potential under disbonded coating. Chin and Sabde [8] measured current and potential distributions on the steel beneath a simulated disbonded coating with a holiday when the holiday potential was controlled at a value in the range of −0.85 to −1.25 V/SCE in 0.001–0.6 M NaCl solutions. They found that the transient potential distribution in the circular crevice became more uniform with increasing time, but the transient current distribution had a reverse trend.

In the present study, potential and current distributions in a cathodically protected crevice between a simulated disbonded coating and segmented steel electrodes were measured in dilute NaCl solutions. The effects of control potential, initial solution conductivity and crevice thickness on the distributions were examined. On the basis of the results, a mechanism of CP against crevice corrosion in high-resistivity environments was proposed.

EXPERIMENTAL

All experiments were carried out using a cell designed to simulate a crevice beneath a disbonded coating, as shown in Fig. 1(a) and (b). A rectangular crevice 94 mm in length and 19 mm in width was formed by bolting a polytetrafluoroethylene (PTFE) gasket between two polymethyl methacrylate (PMMA) plates. The thickness of the PTFE gasket was chosen to control the crevice thickness. A hole 74 mm in diameter was drilled at one end of the top plate and a PMMA cylinder (74 mm in inside diameter, 47 mm in height and 3 mm in wall thickness) was glued to the top end of the hole to serve as the bulk solution compartment. One end of the crevice opened to the bottom edge of the bulk solution compartment to serve as a holiday opening to the crevice. A rectangular titanium sheet ($100 \times 25 \times 2$ mm^3) served as the counter electrode and was put in the bulk solution parallelly to the holiday opening. A Luggin capillary connected to a SCE was put at the center of the holiday opening. Eight mild steel electrodes were sealed and isolated each other in the rectangular groove along the crevice length. They were connected respectively to the work electrode output of a potentiostat through parallel conducting wires to serve as the cathode. The work surface (10×15 mm^2) of every electrode and the upper surface of the bottom plate were located on the same plane to keep a uniform crevice thickness. The width of every electrode along the crevice length was 10 mm and the width of an inert section between two electrodes was 2 mm. The electrode nearest to the holiday opening was labeled 1 and the one furthest from the holiday opening was labeled 8.

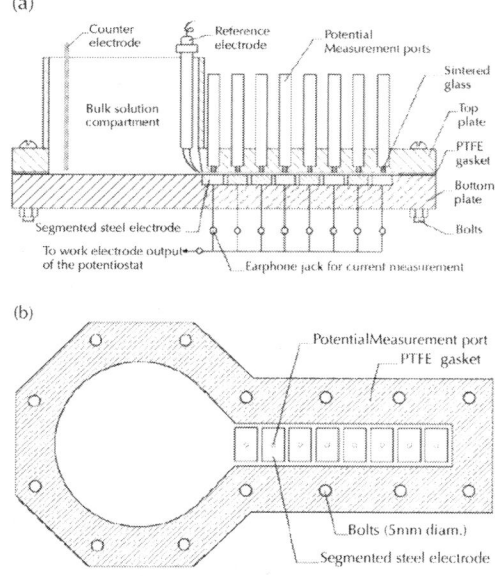

Figure 1: (a) Schematic of the model crevice cell. (b) Top view of the crevice cell showing the locations of the segmented steel electrodes and potential measurement ports.

In order to measure the potential distribution in the crevice, eight potential measurement ports were installed. Above the center of every steel electrode a hole was drilled through the top PMMA plate. Then its bottom end 2 mm in diameter was filled with sintered glass and its top end 8 mm in diameter was fitted with a PMMA tube 6 mm in inside diameter to serve as a potential measurement port. The distances from the holiday opening to these ports were 5, 17, 29, 41, 53, 65, 77 and 89 mm according to priority. Putting another SCE into these ports, the local potential on the steel in the crevice (E) can be measured with a digital multimeter one by one.

An earphone jack was connected into the parallel conducting wire of every steel electrode so that a KEITHLEY 485 picoammeter can easily connected into the electric circuits with an earphone plug to measure the current through every electrode. The current density at the center point of every electrode (I) was calculated by dividing the measured current with the area of one steel electrode (1.5 cm^2).

Prior to each experiment, the work surface of every steel electrode was polished using 0–3$^\#$ emery paper, washed with distilled water, rinsed with absolute ethanol and dried with acetone. After the crevice cell was assembled, the cell and all the potential measurement ports were filled with a test NaCl solution. The potential of the steel cathode was controlled at a constant value in the range of −0.95 V to −1.25 V with respect to the SCE located at the center of the holiday opening using the potentiostat. The local potential (E) and current density (I) on the steel in the crevice were recorded at various time intervals. The experimental duration lasted for 25 h except for that at a control potential of −1.25 V/SCE because hydrogen bubbles would block the Luggin capillary. At the end of an experiment, the solution in the crevice was taken out for measuring average conductivity and pH, and corrosion damages on the steel surface was observed visually.

The test NaCl solutions, with concentrations of 0.06, 0.006 and 0.0006 M (pH$\cong$6.6, at 25°C), were prepared from reagent grade NaCl and double distilled water. All experiments were performed at 25±1°C.

RESULTS

A set of CP experiments against crevice corrosion in the crevice cell were carried out to examine the effects of control potential at holiday (E_h), initial solution conductivity ($\sigma_{initial}$) and crevice thickness (δ) on the potential and current distributions. Some results of these experiments including final average solution conductivity (σ_{final}) and pH (pH$_{final}$) in the crevice as well as corrosion damage on steel were listed in Table 1. Only slight or very slight corrosion was observed on the steel surface. The solution pH in the crevice increased from an initial value of 6.6 to a final value of 8.6–10.6, and the solution conductivity in the crevice increased remarkably, especially for the experiments in 0.006 and 0.0006 M NaCl solutions. These changes were caused by the generation of hydroxyl ions (OH^-) from the oxygen reduction reaction in the crevice and by the migration of sodium ions (Na^+) from bulk solution into the crevice.

Table 1: Experimental conditions and results of CP against crevice corrosion on steel

No.	E_h(V/SCE)	C_{NaCl}(M)	δ(mm)	$\sigma_{initial}$(μS/cm)	t_p(h)	σ_{final}(μS/cm)	pH$_{final}$	Corrosion on steel
1	−0.95	0.006	1.0	720	25	1850	8.7	Slight[a]
2	−1.05	0.006	1.0	720	25	2000	9.2	Very slight[b]
3	−1.15	0.006	1.0	720	25	1750	9.5	Very slight
4	−1.25	0.006	1.0	720	13	1350	9.7	Very slight
5	−1.05	0.06	1.0	6500	25	7400	9.8	No
6	−1.05	0.0006	1.0	115	25	560	8.6	Slight
7	−1.05	0.006	0.5	720	25	1650	9.8	Slight
8	−1.05	0.006	1.5	720	25	1850	10.6	Very slight

[a]Slight corrosion – about 40% of the steel surface was covered with a light brownish and blackish film with the rest being bright and lustrous.

[b]Very slight corrosion – only about 10% of the steel surface was covered with a light brownish and blackish film with the rest being bright and lustrous.

Changes of Potential and Current Distributions with Time

Fig. 2 and Fig. 3 show respectively potential and current distributions (E–x and I–x) in the crevice recorded at various experimental times in 0.006 M NaCl solution at a control potential of −1.15 V/SCE. As shown in Fig. 2, a significant potential gradient existed in the crevice. The magnitude of the gradient decreased with time. At the initial stage, the current concentrated on the steel surface near the holiday, and current density on the steel surface in deeper areas of the crevice was much smaller. Moreover, an obvious anodic current appeared on electrode 6. The anodic current was generated by local corrosion cell, which formed between adjacent steel electrodes due to the heterogeneity of the steel material and chemical environment. However, as the experimental time prolonged, the total cathodic current along with the local current in the front of the crevice decreased observably, the local current near the bottom of the crevice increased gradually, and the anodic current disappeared. After 25 h, the potential and current distributions in the crevice tended to become stable. These results indicated that the potential and current distributions in the crevice became more uniform with increasing time. The disappearance of the anodic current implied that CP might prevent the local corrosion in the crevice.

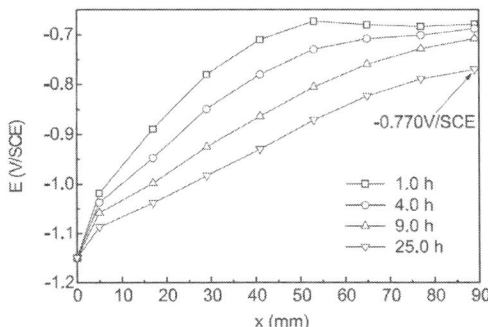

Figure 2: Potential distributions in the crevice at various experimental times in 0.006 M NaCl (E_h=−1.15 V/SCE, δ=1 mm).

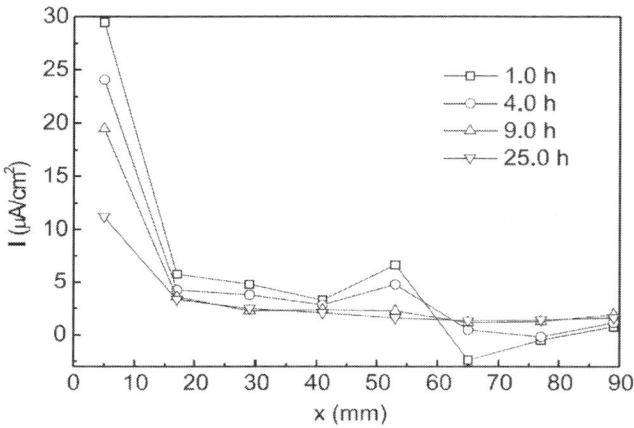

Figure 3: Current distributions in the crevice at various experimental times in 0.006 M NaCl (E_h=−1.15 V/SCE, δ=1 mm).

Effect of Control Potential on Potential and Current Distributions

After the steel electrodes were polarized for 11 h at the control potentials of −0.95, −1.05, −1.15 and −1.25 V/SCE, potential and current distributions in a 0.006 M NaCl solution were plotted respectively in Fig. 4 and Fig. 5. In general, a more negative control potential promoted a higher cathodic polarization level in the crevice. However, the local potential near the crevice bottom at a control potential of −1.25 V/SCE was less negative than that at a control potential of −1.15 V/SCE. This abnormal phenomenon was caused by the hydrogen bubbles produced at the mouth of the crevice [12]. In fact, at the control potential of −1.25 V/SCE, the hydrogen evolution on the steel surface in the immediate vicinity of the holiday was so vigorous that the hydrogen bubbles blocked the flow of current into the crevice. The current density near the holiday increased as the control potential moved in the more negative direction. However, the control potential had little influence on the current distribution in deeper regions of the crevice due to the low solution conductivity.

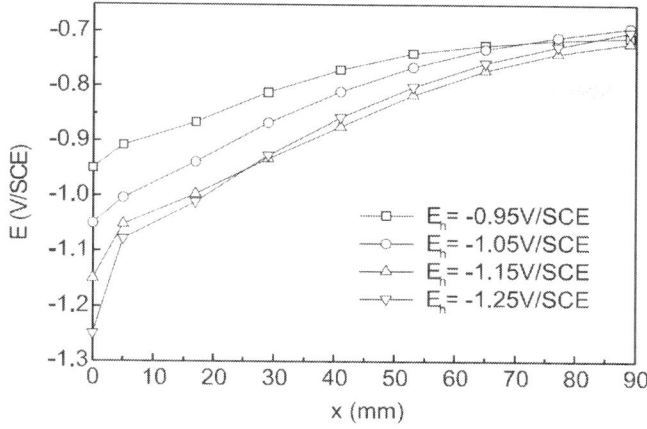

Figure 4: Potential distributions in the crevice at 11 h in 0.006 M NaCl at four different control potentials (δ=1 mm).

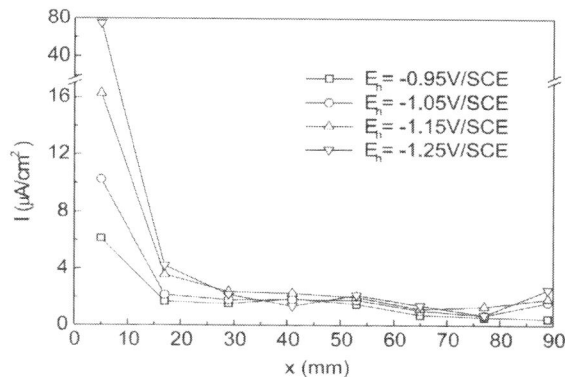

Figure 5: Current distributions in the crevice at 11 h in 0.006 M NaCl at four different control potentials (δ=1 mm).

Effect of Initial Solution Conductivity on Potential and Current Distributions

The effect of initial solution conductivity on potential and current distributions in the crevice was examined in 0.06, 0.006 and

0.0006 M NaCl solutions at a control potential of −1.05 V/SCE, as shown in Fig. 6 and Fig. 7. At the initial stage (0.15 h), the potential gradient decreased and the local current density increased with an increase in initial solution conductivity. However, when initial solution conductivity was less than 720 μS/cm the potential distributions at 25 h (Fig. 6) had no significant difference. In the meantime, initial solution conductivity had no important effect on the current distributions at 25 h (Fig. 7) due to the depletion of dissolved oxygen and an increase of solution conductivity in the crevice.

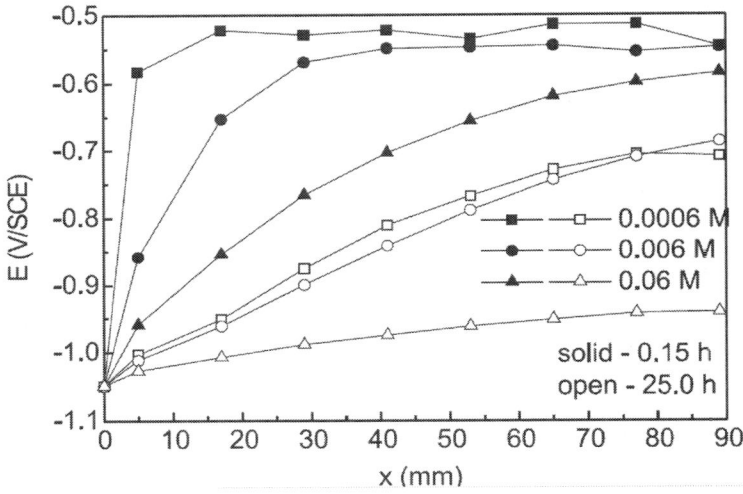

Figure 6: Potential distributions in the crevice with three different NaCl concentrations (E_h=−1.05 V/SCE, δ=1 mm).

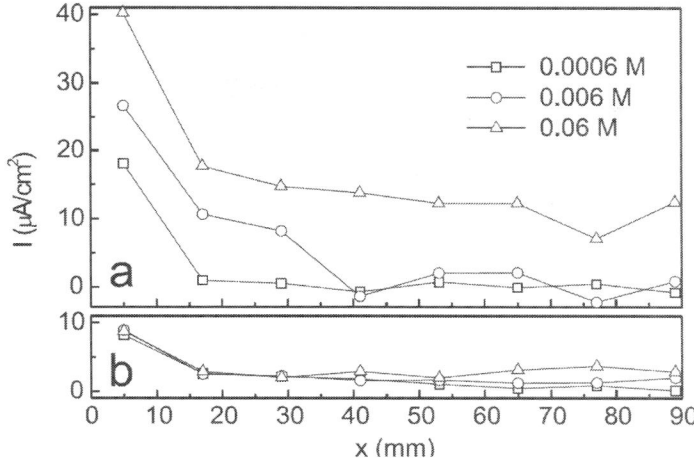

Figure 7: Current distributions in the crevice with three different NaCl concentrations at (a) 0.15 h and (b) 25 h ($E_h=-1.05$ V/SCE, $\delta=1$ mm).

In a crevice filled with a higher-conductivity solution (such as 0.06 M NaCl solution), the solution ohmic resistance is small, and the rate of cathodic reaction on the steel was determined by the concentration of dissolved oxygen in the crevice. Therefore, the local current density on the steel was large at the initial stage of polarization and then decreased obviously with time due to the consumption of dissolved oxygen in the crevice. However, in a crevice filled with a lower-conductivity solution the rate of cathodic reaction on steel was determined by the solution ohmic resistance because of a big IR drop with the exception in the vicinity of the holiday. Therefore, the local current density on the steel was very small at the initial stage of polarization and then increased gradually with time due to an increase in conductivity of the crevice solution.

Effect of Crevice Thickness on the Potential and Current Distributions

Fig. 8 and Fig. 9show respectively the potential and current distributions in the crevices with three different thicknesses (i.e.,

0.5, 1.0 and 1.5 mm) in 0.006 M NaCl solution after the steel electrodes were polarized for 25 h at a control potentials of −1.05 V/SCE. Generally, the potential gradient decreased and the local current density increased with an increase in crevice thickness. However, the effect of the crevice thickness from 0.5 to 1.0 mm on the potential and current distributions was less significant than that from 1.0 to 1.5 mm. Moreover, the potential and current distribution curves of 0.5 mm crevice thickness crossed those of 1.0 mm crevice thickness at some distance, which may be caused by the more durative effect of the local corrosion cell in the crevice with 0.5 mm thickness.

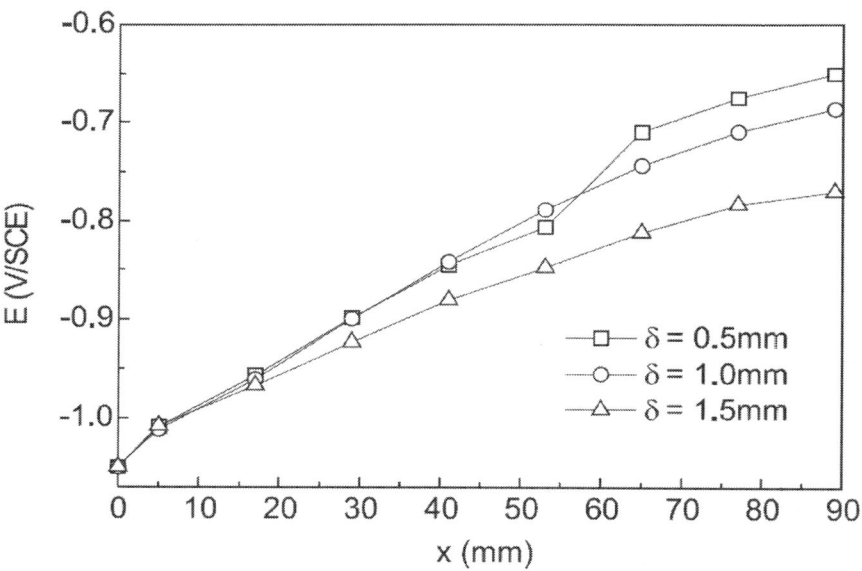

Figure 8: Potential distributions in the crevice with three different crevice thicknesses in 0.006 M NaCl at 25 h (E_h=−1.05 V/SCE).

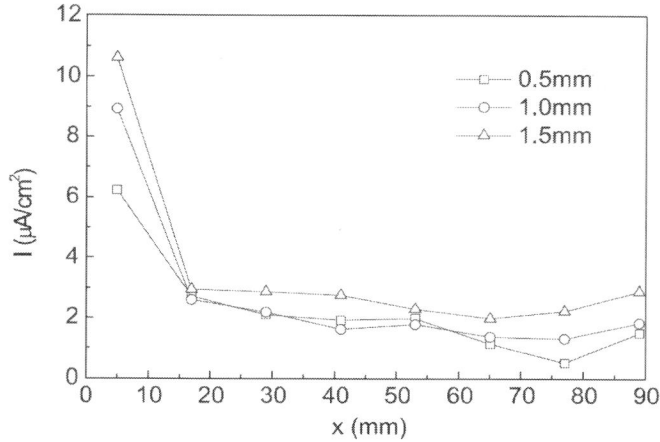

Figure 9: Current distributions in the crevice with three different crevice thicknesses in 0.006 M NaCl at 25 h (E_h=−1.05 V/SCE).

DISCUSSION

Mathematical Analysis of the Potential and Current Distributions

Because the simulated crevice used in the present work is a rectangular crevice and the crevice depth is much larger than the crevice thickness, local potential and current density in the crevice can be considered approximately as one-dimensional variables in the depth direction (x). On the assumption that the current flow in the crevice solution satisfies Ohm's law and the solution conductivity ( ) was constant, the relation between the local potential and current density on the steel in the crevice can be described by a special type of the Laplace equation [13]:

$$\frac{\mathrm{d}^2 E(x)}{\mathrm{d}x^2} = -\frac{1}{\sigma\delta} I(x)$$

(1)

Because the local current density on the steel electrodes in the crevice was typically in the order of 1 $\mu A/cm^2$, the following linear polarization equation may be assumed.

$$E(x) = E_{corr} - R_p I(x) \tag{2}$$

Where E_{corr} is the corrosion potential and R_p is the linear polarization resistance for cathodic reaction. If boundary conditions are $E(x) = E_0$ at $x=0$ and $E(x) = E_{corr}$ at $x \to \infty$, and can be combined to obtain the potential distribution $E(x)$:

$$E(x) = E_{corr} + (E_0 - E_{corr}) \exp(-x/a) \tag{3}$$

Where

$$a = \sqrt{R_p \sigma \delta} \tag{4}$$

When Eq. (2) was substituted into Eq. (3), the current distribution $I(x)$ resulted

$$I(x) = \frac{E_{corr} - E_0}{R_p} \exp(-x/a) \tag{5}$$

In order to examine the accuracy of this mathematical model, and were fitted to the experimental data obtained under various experimental conditions and at given times using a nonlinear least-squares program. Fig. 10 and Fig. 11present some typical fitting results. This simple model is based upon the Ohm's law and does not consider the transport processes, and the behavior of potential distribution mainly depend on the ohmic potential drop in the crevice, so Eq. (3) agree quite well with the experimental potential distributions as shown in Fig. 10. However, the behavior of current distribution is determined not only by the ohmic resistance effects but also by the transport processes in the crevice, and therefore Eq. (5) can only describe the general trend of the experimental current distributions as shown in Fig. 11. To describe the current distribution more accurately, the transport processes in the crevice must be involved in a mathematical model.

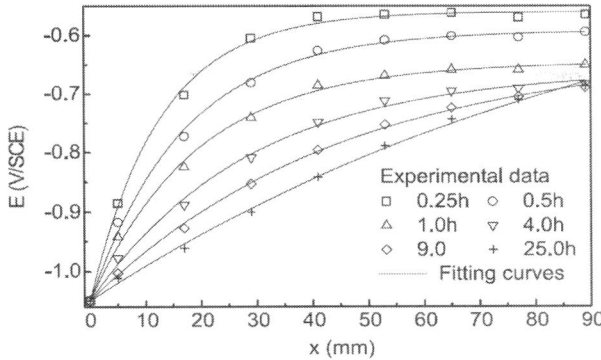

Figure 10: Experimental E–x data in 0.006 M NaCl at various times and the fitting curves of Eq. (3) to the data (E_h=−1.05 V/SCE, δ=1 mm).

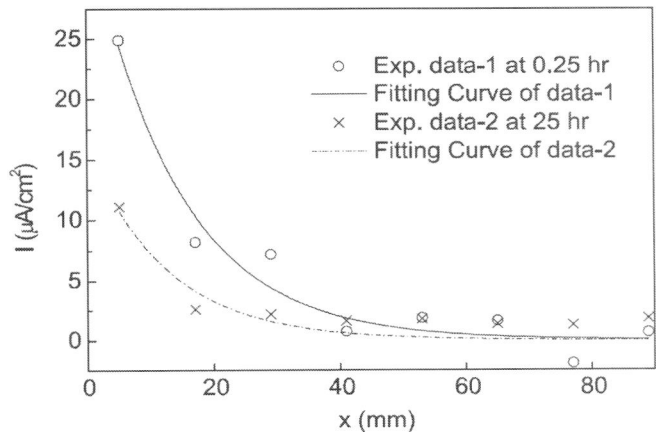

Figure 11: Experimental I–x data in 0.006 M NaCl at various times and the fitting curves of Eq. (5) to the data (E_h=−1.05 V/SCE, δ=1 mm).

According to Eq. (3), the average potential gradient in the crevice can be shown as

$$\frac{\Delta E}{\Delta x} = \frac{E_{\text{corr}} - E_0}{L}\left[1 - \exp\left(-L/a\right)\right]$$

(6)

Where L is the length of the crevice.

Because potential and current distributions in the crevice became more uniform with a larger value of the parameter a, we defined a as "uniform index". In a cathodically protected crevice containing a dilute NaCl solution, hydroxyl ions (OH⁻) are produced by oxygen reduction and the excess OH⁻ ions in turn attract Na⁺ in the bulk solution to migrate into the crevice. As a result, the conductivity of the crevice solution increased. Moreover, a decrease in the concentration of dissolved oxygen can result in a significant increase in R_p [14]. Therefore, the uniform index increase with increasing time. In Fig. 12, a curve of a vs. time was calculated from the potential distribution data and plotted. Indeed, the uniform index increased with time. On the other hand, according to , , and , if the initial solution conductivity and the crevice thickness increase, the potential gradient decrease and the local current increase in the crevice. Moreover, a more negative control potential causes a higher polarization level and larger current density on the steel in the crevice. These conclusions are corresponding to the experimental results.

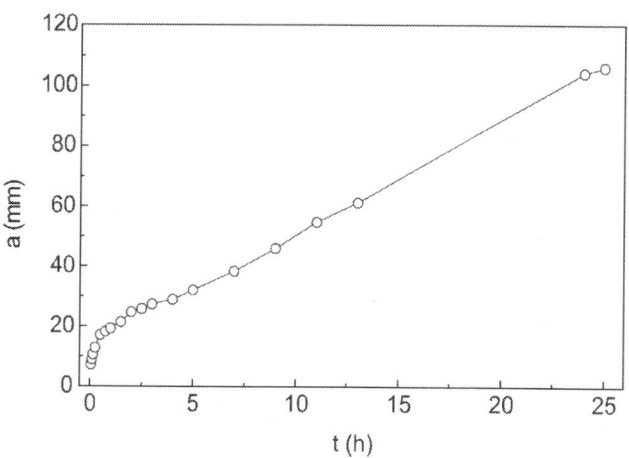

Figure 12: Curve of uniform index (a) vs. time in 0.006 M NaCl (E_h=−1.05 V/SCE, δ=1 mm).

Mechanism of Cathodic Protection against Crevice Corrosion in High-resistivity Environments

In high-resistivity environments, although it is difficult for cathodic current to penetrate deeply into the crevice, the corrosion in the crevice can be controlled effectively at a proper control potential. Based upon the changes of potential and current distributions as well as the change of chemical environment in the cathodically protected crevice containing a dilute NaCl solution, a "three-stage complex mechanism" was proposed for the process of CP against crevice corrosion:

- At initial stage of the process, only the steel surface in the vicinity of a crevice opening falls under direct electrochemical protection and corrosion still occurs in deeper areas of the crevice due to the existence of dissolved oxygen. This stage can be called as the "finite electrochemical protection stage".

- When the process lasts for a period of time, the dissolved oxygen originally in the crevice is depleted by the cathodic current, the corrosion in deeper areas of the crevice stops, and then the process come into the second stage – the "oxygen depletion stage". This is proved by the fact that the local potential in the crevice shifted to negative direction with increasing time, but meanwhile the total cathodic current decreased with increasing time. At this stage, the steel surface in the vicinity of a crevice opening still falls under direct electrochemical protection.

- When the process lasts for enough long time, the chemical environment in the crevice changes greatly. The protection current may penetrates into deeper areas of the crevice due to a significant increase in conductivity of the crevice solution and make the entire steel surface in the crevice fall under direct electrochemical protection. Then the process enters into the "complete electrochemical protection stage". In some conditions, the experimental result that final local potential at the bottom of the crevice at 25 h approached the CP criterion (see Fig. 2) may be explained by this mechanism.

CONCLUSIONS

From above the experimental results and discussion, the following conclusions are drawn with respect to CP against crevice corrosion in the present test conditions:

- The pH and conductivity increased in the cathodically protected crevice because of the generation of hydroxyl ions from the oxygen reduction in the crevice.
- The potential and current distributions in the crevice became more uniform with increasing time due to an increase in solution conductivity and depletion of dissolved oxygen in the crevice.
- As the initial solution conductivity and the crevice thickness increased, the potential gradient decreased and local current increased in the crevice.
- A more negative control potential resulted in a higher polarization level and larger current density on the crevice steel. However, if control potential is too negative, the cathodic polarization of the steel near the crevice bottom may be less than that at a suitable control potential because of hydrogen evolution.
- The process of CP against crevice corrosion in high-resistivity environments can be divided into the three stages, a "three-stage complex mechanism" was proposed.

ACKNOWLEDGMENTS

This research was supported by the National Science foundation of China and the Chinese State Key Laboratory for Corrosion and Protection of Metals.

REFERENCES

1.   G Mill Materials Performance, 27 (12) (1988), p. 13
2.   J.A Beavers, N.G Thompson Materials Performance, 36 (4) (1997), p. 13
3.   A.C Toncre Materials Performance, 23 (8) (1984), p. 22
4.   A.C Toncre, N Ahmad Materials Performance, 19 (6) (1980), p. 39
5.   M.D Orton Materials Performance, 24 (6) (1985), p. 17
6.   M.H Peterson, T.J Lennox Jr. Corrosion, 29 (10) (1973), p. 406
7.   G Sabde, F Gan, D.T Chin Journal of Chinese Institution of Chemical Engineering, 24 (6) (1993), p. 417
8.   D.-T Chin, G.M Sabde Corrosion, 55 (3) (1999), p. 229
9.   JF Gan, Z.W Sun, G Sabde, D.T Chin Corrosion, 50 (10) (1994), p. 804
10.   T.R Jack, G.V Boven, M Wilmott, R.L Sutherby, R.G Worthingham Materials Performance, 33 (8) (1994), p. 17
11.   R Brousseau, S Qian Corrosion, 50 (12) (1994), p. 907
12.   R.R Fessler, A.J Markworth, R.N Parkins Corrosion, 39 (1) (1983), p. 20
13.   W Schwenk Corros. Sci., 23 (8) (1983), p. 871
14.   Z Li, X Mao, F Gan Journal of Chinese Society for Corrosion and Protection, 20 (3) (2000), p. 129 (in Chinese)

Chapter 6

Improvement in Corrosion Inhibition Efficiency of Molybdate-Based Inhibitors via Addition of Nitroethane and Zinc in Stimulated Cooling Water

Saeed Mohammadi, Fatemeh Baghaei Ravari, and Athareh Dadgarinezhad

Department of Materials Science and Engineering, Faculty of Engineering, Shahid Bahonar University, Kerman 7618868366, Iran

ABSTRACT

An investigation was conducted to improve the corrosion inhibition efficiency of molybdate-based inhibitors for mild steel which is the main construction material of cooling water systems, using nitroethane as an organic compound beside zinc. In this study a new molybdate-based inhibitor was introduced with the composition of 60 ppm molybdate, 20 ppm nitrite, 20 ppm nitroethane, and 10 ppm zinc. Inhibition efficiency of molybdate alone and with nitrite, nitroethane, and zinc on the uniform corrosion of mild steel in stimulated cooling water (SCW) was assessed by electrochemical techniques such as potentiodynamic polarization and electrochemical impedance (AC impedance) measurements. Weight loss measurements were made with coupon testing specimens in the room temperature for 48 h. Studies of electron microscopy, including scanning electron microscopy (SEM) photograph and X-ray energy dispersive spectrometry (EDS) microanalysis, were used. The results obtained from the polarization and AC impedance curves were in agreement with those from the corrosion weight loss results. The results indicate that the new inhibitor is as effective as molybdate alone, though at one-ninth of the concentration range of molybdate, which is economically favorable.

INTRODUCTION

The protection of cooling water systems as well as heat supply water has become one of the great important issues in the world economy. The application of corrosion inhibitors especially in closed systems holds a prominent place amongst other methods of corrosion control [1]. The actual trends in the environmental protection essentially have changed the traditional approach to corrosion inhibition. Since the toxicity of chromate-based inhibitor is a limiting factor in its use as a corrosion inhibitor, the changes in formulation of corrosion inhibitors are prompted primarily by an increasing demand to reduce environmental impact [2]. Molybdate-based inhibitor has long been known as an inorganic and anodic type of corrosion inhibitor, which is effective for protecting mild steel in the pH range 5.8–8.5 [3, 4]. Lizlovs has observed that in the aqueous system containing aggressive ions, molybdate has corrosion inhibition only in the presence of oxygen

[5]. In fact, the presence of aggressive ions such as chloride (Cl^-) and sulfate (SO_4^{-2}) anions reduces the efficiency of MoO_4^{2-}, so higher concentrations are necessary for corrosion inhibition [6, 7], which is not economically favorable. In order to achieve better efficiency and reduce the quantity of molybdate, other oxidizing agents such as nitrite (NO_2^-) and organic compounds have been employed. Recently, the best method to improve inhibitive capability is using inhibitors in combination with others [8–11]. As it has been observed previously, organic inhibitors usually designated as a film forming protect the metal by forming a hydrophobic film on the metal surface. Therefore, natural organic molecules containing one pair of electrons or associated with multiple specially triple bonds or organic rings can bond to metal surface by electron transfer to the metal to form a coordinate type of link, which ultimately produces a barrier to the dissolution of the metal in the electrolyte [12]. As nitroethane contains a carbon chain together with oxidizing agent (NO_2^-), it is predicted that it can improve inhibition efficiency as an organic compound. In addition, Jefferies and Bucher recently studied the addition of zinc as cathodic inhibitor to improve the corrosion inhibition behavior of molybdate-based inhibitors [13]. As a result, it is worth investigating the corrosion inhibition of mild steel in stimulated cooling water (SCW) by using new developed inhibitor containing molybdate, nitrite, nitroethane, and zinc.

EXPERIMENTAL PROCEDURES

Coupon testing specimens with the dimensions of 5, 2, and 0.3 cm were used for weight loss measurements. The composition of mild steel specimens is shown in Table 1.

Table 1: Chemical composition of test specimen

Element	C	Si	Mn	Fe
Test specimen	0.05	0.49	0.54	98.92

Mild steel of the same alloy composition with an exposed area of 1 cm² was embedded in epoxy resin and used for electrochemical measurements. Tafel polarization measurements were carried out at open circuit potential (E_{ocp}), potentiostat/galvanostat (Princeton Applied Research EG & G Model 263 A), using counter electrode (Pt) and a saturated calomel electrode (SCE) as reference electrode. All the quoted potentials are referred to this reference electrode. The potentiodynamic current-potential curves were carried out at a scan rate 1 mV/sec. For impedance measurement, the same equipment was used for the Tafel polarization measurement and combined with frequency response analyzer (Princeton Model 1020). The impedance measurement was carried out using AC signals of amplitude 10 mV peak to peak at the open circuit potential in the frequency range 100 KHz to 10 mHz. Before each test, the specimens were prepared with silicon carbide paper 220–1200 grit, degreased in acetone, and washed in distilled water. Test solutions with different inhibitors (Table 2) were prepared by dissolving analytical grade sodium molybdate (Na_2MoO_4), sodium nitrite ($NaNO_2$), zinc sulfate ($ZnSO_4$), and sodium nitroethane ($C_2H_5NNaO_2$).

Table 2: Chemical composition of inhibitors

Inhibitor	Concentrations (ppm)
1000M	1000 Molybdate sodium
40N	40 Nitrite sodium
20Ne	20 Nitroethane sodium
60M40N	60 Molybdate sodium + 40 nitrite sodium
40N20Ne	40 Nitrite sodium + 20 nitroethane sodium
60M20Ne	60 Molybdate sodium + 20 nitroethane sodium
60M40N10Zn	60 Molybdate sodium + 40 nitrite sodium + 10 zinc sulfate
60M20Ne10Zn	60 Molybdate sodium + 20 nitroethane sodium + 10 zinc sulfate

60M20N20Ne10Zn	60 Molybdate sodium + 20 nitrite sodium + 20 nitroethane sodium + 10 zinc sulfate

Figure 1 shows the molecular structure of sodium nitroethane. Distilled water with 500 ppm sodium chloride, 520 ppm sodium sulfate, 170 ppm sodium bicarbonate, and 25 ppm sodium carbonate was used as SCW [14].

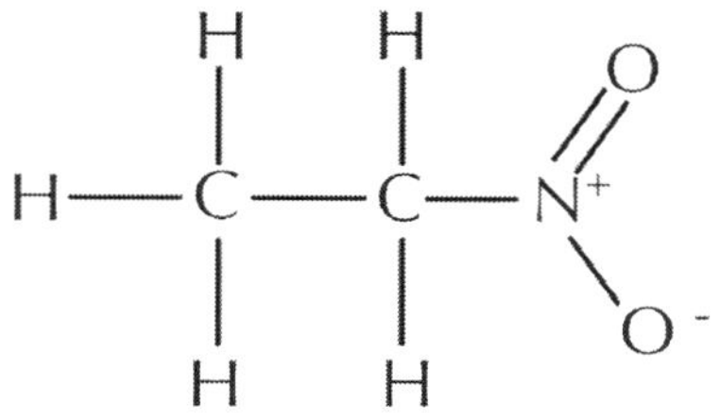

Figure 1: Molecular structure of sodium nitroethane.

All tests were performed at neutral pH and room temperature. Before measurements of polarization curves, a stabilization period of 20 min was observed which proved sufficient as indicated by open circuit potential (E_{ocp}). Table 2 shows the composition of different inhibitors used in this work.

The surface morphology of the mild steel samples after immersion in SCW without inhibitor and with inhibitor 1000 M and the new optimized inhibitor was photographed using SEM and was analyzed by EDS (Cam Scan 2300 MV).

RESULTS AND DISCUSSION

Polarization Curves

The anodic and cathodic polarization curves for mild steel in SCW without inhibitor and with different inhibitors are shown in Figures 2–5. Figure 2 shows polarization curves for mild steel without inhibitor and with inhibitor 1000 ppm molybdate, 40 ppm nitrite, and 20 ppm nitroethane.

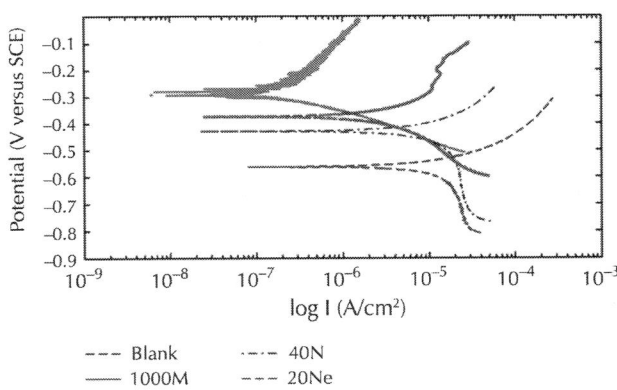

Figure 2: Polarization curves for mild steel in SCW without inhibitor and with inhibitor 1000 ppm molybdate, 40 ppm nitrite, and 20 ppm nitroethane.

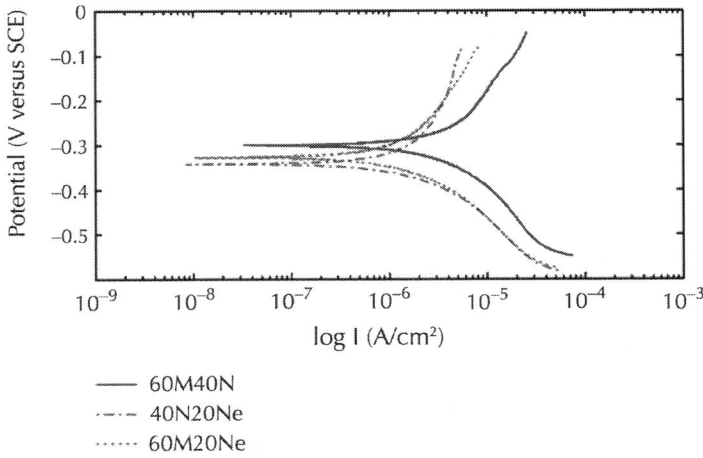

Figure 3: Polarization curves for mild steel exposed to mixture of 60 ppm

molybdate and 40 ppm nitrite, 40 ppm nitrite and 20 ppm nitroethane, and 60 ppm molybdate and 20 ppm nitroethane.

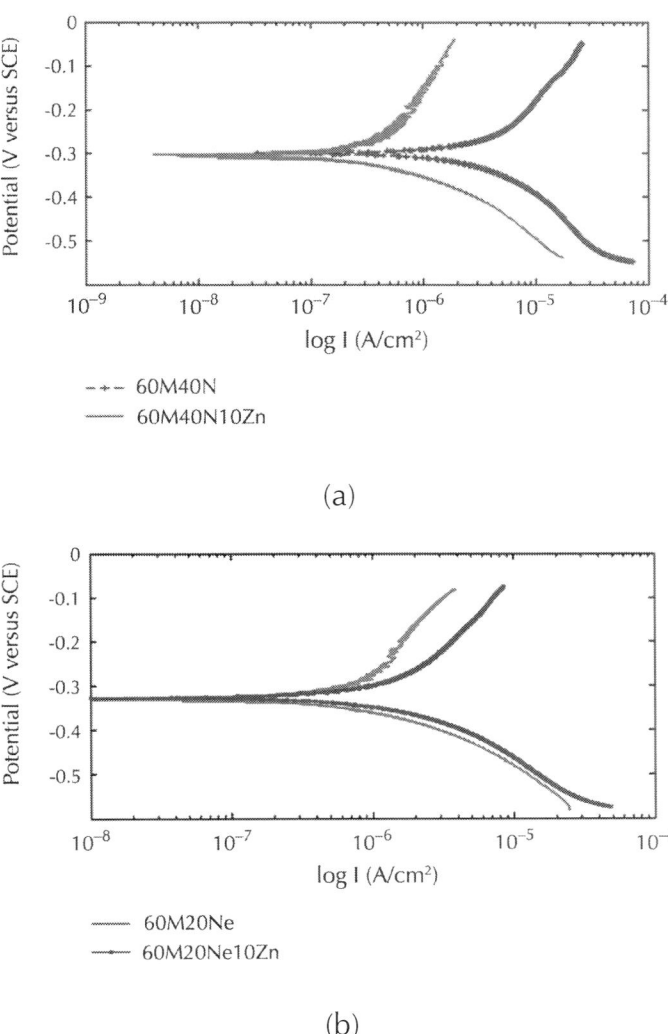

(a)

(b)

Figure 4: Polarization curves for mild steel in SCW by adding zinc to (a) 60 ppm molybdate and 40 ppm nitrite, curve 60M40N10Zn and (b) 40 ppm nitrite and 20 ppm nitroethane, curve 40N20Ne10Zn.

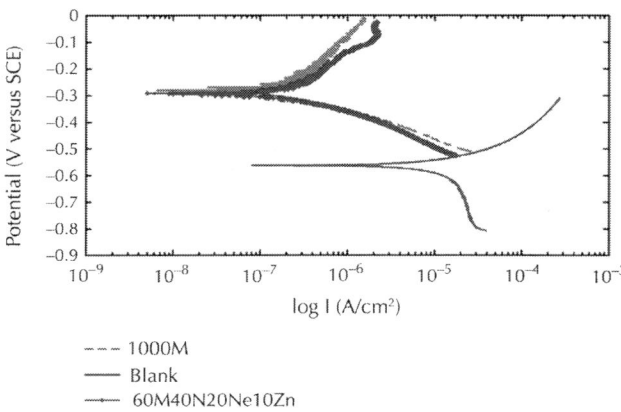

- - - 1000M
—— Blank
·—·— 60M40N20Ne10Zn

Figure 5: Polarization curves for mild steel in SCW without inhibitor and with inhibitor 1000 M and new optimized inhibitor.

The addition of 1000 ppm molybdate, 40 ppm nitrite, and 20 ppm nitroethane shifted the corrosion potentials of the mild steel to more positive values and also decreased the corrosion current densities (I_{corr}), indicating that molybdate, nitrite, and nitroethane are anodic inhibitors.

Table 3 gives values of corrosion potentials (E_{corr}), corrosion current densities (I_{corr}), Tafel slopes (βc and βa), and inhibition efficiency (η%) obtained from polarization measurements of mild steel for different inhibitors in SCW. The corrosion current densities were obtained from the polarization curves by linear extrapolation of Tafel curves at point of 50mV more positive and 50mV more negative than E_{ocp}, respectively.

Table 3: Electrochemical polarization parameters for mild steel in SCW without inhibitor and with different inhibitors

Inhibitor	E_{corr} (mV)	βc (mV/dcade)	βa (mV/dcade)	I_{corr} (μA/cm²)	η (%)
Blank	−550	135	325	14.5	—
1000M	−285	78	187	0.186	98.6
40N	−471	154	188	5.6	61.4

20Ne	−380	98	103	1.82	87.4
60M40N	−308	141	137	1.63	88.7
40N20Ne	−340	68	74	0.83	94.3
60M20Ne	−330	108	153	0.59	96
60M40N10Zn	−318	103	86	0.7	95.2
60M20Ne10Zn	−336	79	144	0.23	98.2
60M20N20Ne-10Zn	−298	90	196	0.173	98.8

The inhibition efficiency is defined as

$$\eta(\%) = \frac{I_{corr} - I_{corr(i)}}{I_{corr}} \times 100,$$

(1)

where I_{corr} and $I_{corr}(i)$ are the corrosion current density values without inhibitor and with different inhibitors, respectively.

For the curve 1000 M, corrosion current density is 0.185 ($\mu A/cm^2$), indicating a negligible corrosion rate. But it is not suitable because the concentration is high. According to the obtained results, molybdate was considered as a weak oxidizer, so other oxidizing agents such as nitrite and nitroethane were added to molybdate in order to achieve better efficiency and reduce the quantity of molybdate.

Figure 3 exhibits the polarization curves for mild steel exposed to mixture of 60 ppm molybdate and 40 ppm nitrite, 40 ppm nitrite and 20 ppm nitroethane, and 60 ppm molybdate and 20 ppm nitroethane.

According to the curves 60M40N, a concentration of 60 ppm MoO_4^{2-} and 40 ppm NO_2^- appeared to be more efficient than molybdate and nitrite alone at mentioned concentrations [11]. Also by considering curve 40N20Ne, one should point out that in solution containing both nitrite and nitroethane ions, anodic curves were shifted to more positive values compared with nitrite and nitroethane alone, indicating synergistic behavior and improvement in the inhibition efficiency. According to Figure 2, nitroethane with half of the concentration range of nitrite exhibits better inhibition corrosion efficiency. So the combination of 60 ppm molybdate and

20 ppm nitroethane was used in order to achieve better efficiency and reduce the corrosion current densities of Tafel curve 60M40N.

Considering the fact that cathodic inhibitors improve the corrosion inhibition efficiency and reduce the corrosion current densities, combination of zinc with mixture of 60 ppm molybdate and 40 ppm nitrite and mixture of 60 ppm molybdate and 20 ppm nitroethane was used to improve the corrosion inhibition efficiency of Tafel curves 60M40N and 60M20Ne, as shown in Figure 4.

According to the Tafel curves 60M40N10Zn and 60M20Ne10Zn, addition of zinc (with delay to cathodic reactions) to oxidizing agents such as molybdate, nitrite, and nitroethane improves the corrosion inhibition efficiency and reduce the corrosion current densities of mild steel specimens in SCW. Moreover, addition of zinc to Tafel curves 60M40N and 40N20Ne shifts the cathodic and anodic branches of the Tafel plots to less corrosion current values (I_{corr}) at relatively the same corrosion potentials, indicating that zinc acts as cathodic inhibitor besides other anodic inhibitors.

Polarization curves for mild steel were exposed to SCW water by adding molybdate, nitrite, nitroethane, and zinc (60 ppm, 20 ppm, 20 ppm, and 10 ppm, resp.), and 1000 ppm molybdate alone is shown in Figure 5. For comparison polarization curves in SCW and without inhibitor are included.

The suitable combination of $MoO_4^{2-} + NO_2^-$ beside nitroethane and zinc ions provides the necessary oxidizingenvironment to support MoO_4^{2-} film formation and synergically lower the corrosion rate. This combination also overcomes the high concentration of MoO_4^{2-}, with relatively the same E_{corr}, which is economically favorable. Also for optimized inhibitor current density was on the order of 0.173 ($\mu A/cm^2$), which was lowered by about 6.9% compared with 1000 ppm molybdate alone. It can be attributed to the absorption of other ions on the metal in conjunction with MoO_4^{2-} to produce an insoluble compound, providing passivity more readily than MoO_4^{2-} alone [11]. Regarding these results, it can be concluded that the value of corrosion of current density of mild steel in SCW with the addition of molybdate decreases, and with addition of nitrite, nitroethane, and zinc ions to molybdate it decreases more and its inhibition efficiency increases.

AC Impedance Curves

Figures 6 and 7 show the Bode and Bode phase plots of steel electrode without inhibitor and with different inhibitors. The main effect of different inhibitors is an increase in the impedance modulus, $|z|$, below 10 mHz (an increase in polarization resistance (R_p)) and also a higher phase angle.

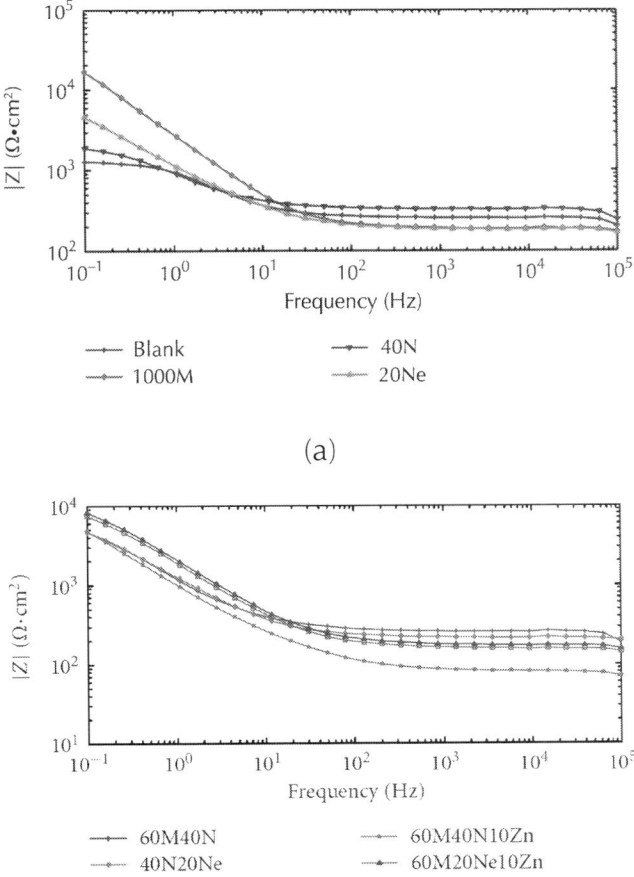

(a)

(b)

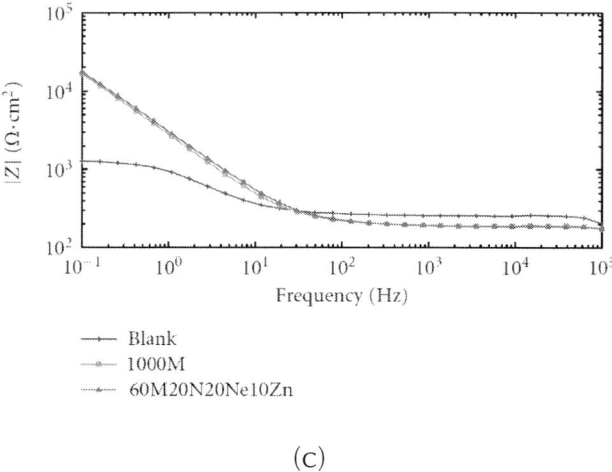

(c)

Figure 6: Bode plots for mild steel in SCW solution (a) without inhibitor and with different inhibitor.

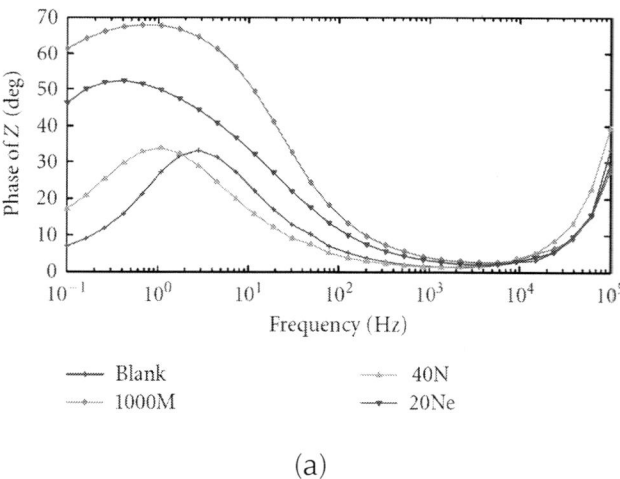

(a)

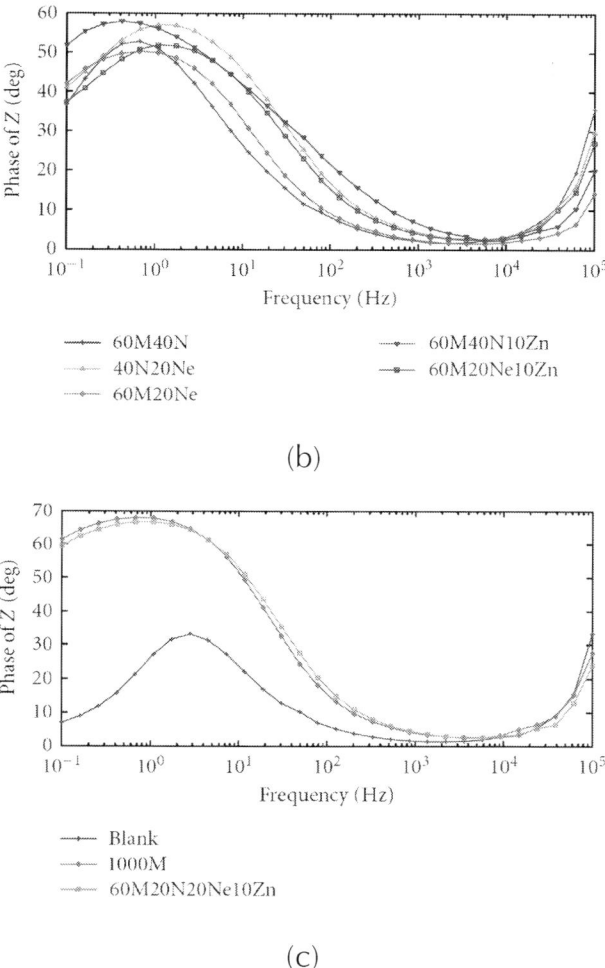

(b)

(c)

Figure 7: Bode phase plots for mild steel in SCW solution without inhibitor and with inhibitor.

In Bode phase plots, only one peak is in the phase angle ( ) versus frequency (f) plot, indicating that there is only one time constant (a single relaxation time constant).

Figure 8 shows Nyquist plots for the steel electrode in SCW solutions without inhibitor and with different inhibitors. The impedance loops measurements were depressed semicircles with their center below the axis. This phenomenon is known as the dispersing effect [15]. In Nyquist plots, the corrosion resistance of each of the samples was determined by R_p. R_p is given by [16]

$$R_p = \lim_{\omega \to 0} R_e \left\{ Z_f \right\}_{E=E_{corr}},$$

(2)

where $Re\{Z_f\}$ represents the real part of the complex faradic impedance, Z_f and ω correspond to the angular velocity of the AC signal ($\omega = 2\pi f$, where f is frequency, (Hz)). R_p values were obtained by fitting the experimental Nyquist data to a simple semicircle and extrapolating to $Z_{im} = 0$.

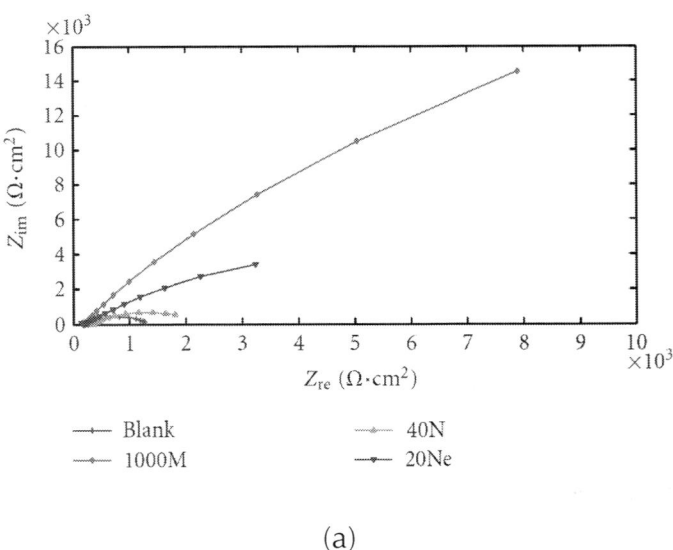

(a)

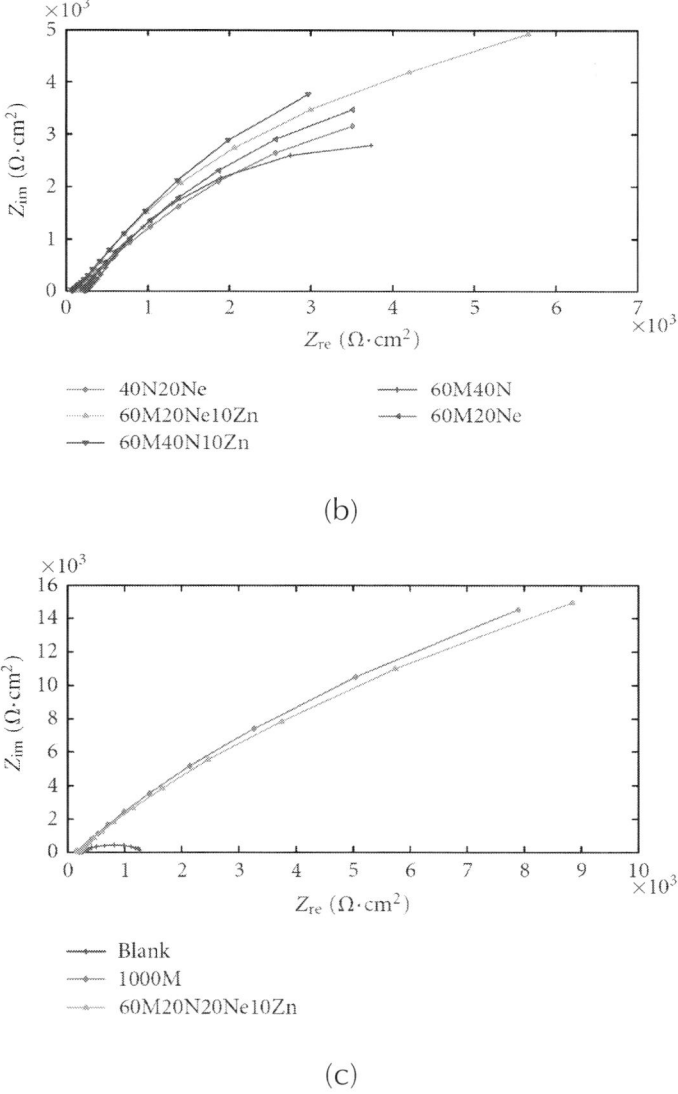

(b)

(c)

Figure 8: Nyquist plots for mild steel in SCW solution without inhibitor and with inhibitor.

By considering AC impedance curves, it was found that polarization resistance increases in the following order.

BLANK < 40N < 20Ne < 40N20Ne ≈ 60M40N < 60M40N10Zn < 60M20Ne < 60M20Ne10Zn < 1000 M ≤ 60M20N20Ne10Zn.

Table 4 gives the values of inhibition efficiency ($\eta\%$) and polarization resistance (R_{ct}) obtained from the AC impedance measurements. The inhibition efficiency is defined as

$$\eta\ (\%) = \frac{R_{ct(i)} - R_{ct}}{R_{ct(i)}} \times 100, \tag{3}$$

where R_{ct} and $R_{ct(i)}$ are the polarization resistance without inhibitor and with different inhibitors, respectively.

Table 4: AC impedance parameters for mild steel in SCW without inhibitor and with different inhibitors

Inhibitor	R_{ct} ($\Omega \cdot cm^2$)	$\eta(\%)$
Blank	1068	—
1000M	16345	93.5
40N	1650	56.8
20Ne	4470	76.1
60M40N	4513	76.6
40N20Ne	4518	76.8
60M20Ne	7083	84.9
60M40N10Zn	4720	77.4
60M20Ne10Zn	7360	85.5
60M20N20Ne10Zn	17205	93.8

The results of the AC impedance test exhibit a good correlation with data obtained from the polarization curves.

Weight Loss Measurements

The weight loss of samples after 48 h exposure to solution SCW without inhibitor and with different inhibitors was measured by coupon testing methods [17]. Regarding experimental weight loss data, inhibition efficiency ($\eta\%$) was calculated classically as follows:

$$\eta(\%) = \frac{W_0 - W_i}{W_0} \times 100,$$

(4)

where W_o and W_i are the weight loss observed in the absence and in the presence of inhibitor, respectively. Values of inhibition efficiency for mild steel in SCW without inhibitor and with different inhibitors are summarized in Table 5.

Table 5: Weight loss measurements for mild steel in SCW without inhibitor and with different inhibitors.

Inhibitor	$\eta(\%)$
Blank	—
1000M	86.5
40N	49.7
20Ne	69.2
60M40N	70.1
40N20Ne	69.8
60M20Ne	78
60M40N10Zn	70.6
60M20Ne10Zn	78.8
60M20N20Ne10Zn	86.8

For comparison, the inhibition efficiency of polarization, AC impedance, and weight loss results were summarized in Figure 9. According to Figure 9, the results obtained from the polarization and AC impedance curves are in agreement with those from the corrosion weight loss tests.

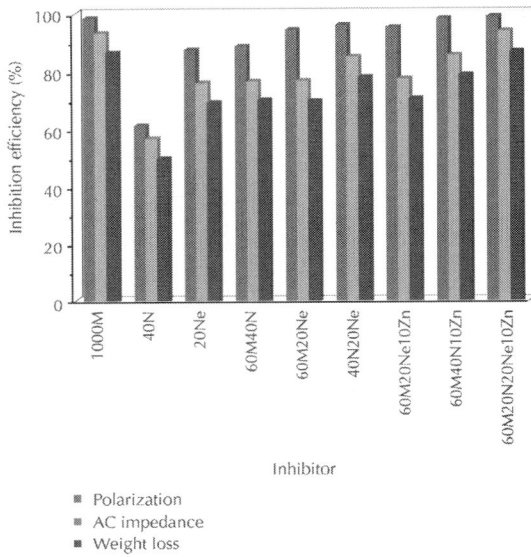

Figure 9: The inhibition efficiency of polarization, AC impedance, and weight loss results for mild steel in SCW with different inhibitors.

Results of Surface Analysis Techniques

SEM Analysis

The presence of corrosion inhibitor could be more clearly investigated by means of surface analysis techniques. SEM micrographs of samples after 48 h exposure to (a) solution SCW without inhibitor, (b) solution SCW with inhibitor 1000 M, and (c) solution SCW with new optimized inhibitor are shown in Figure 10. Figures 10(b) and 10(c) exhibit a layer with uniform, adherent, and continuous structure, while Figure 10(a) shows crystal growth of iron oxide on the surface and its noncontinuous structure. In fact, the decrease in corrosion current densities and increase in polarization resistance when exposed to the previous inhibitors were due to the coverage of metal surface with more protective films [18] as shown in Figure 10. Also the surface roughness of mild steel exposed to solution SCW is much higher than mild steel exposed to solution SCW with the inhibitor 1000 M and optimized inhibitor.

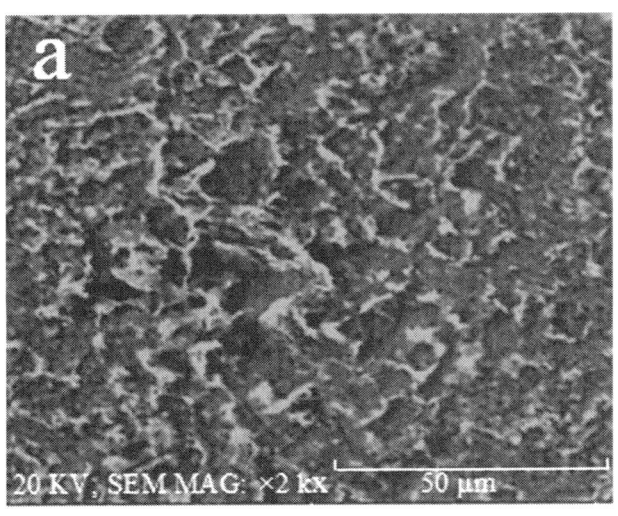

(a)

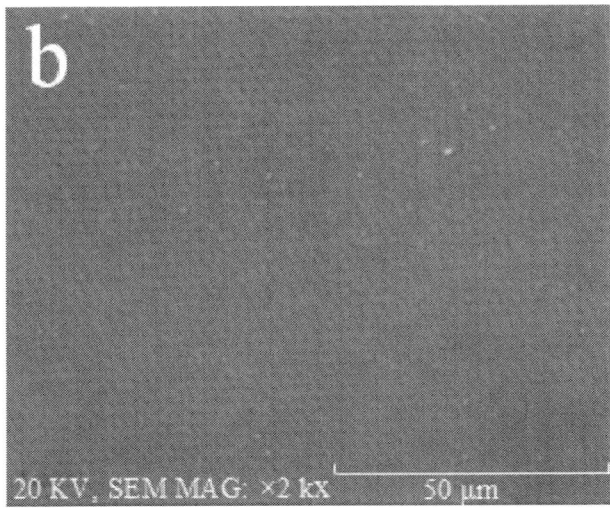

(b)

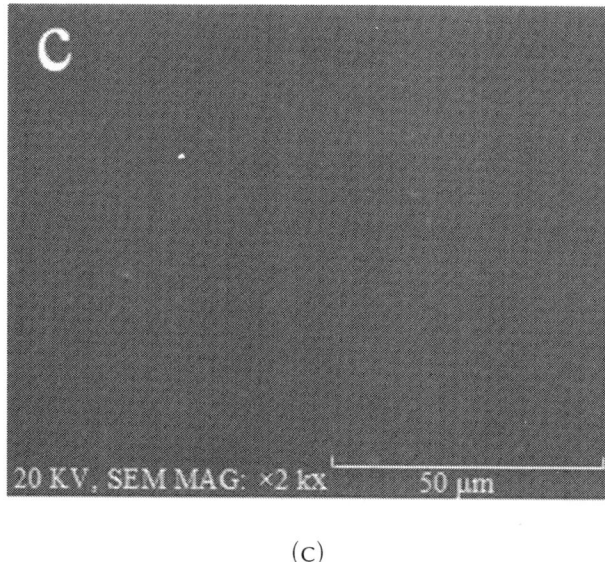

(c)

Figure 10: SEM micrograph for mild steel surface (a) in SCW without inhibitor, (b) in SCW with the inhibitor 1000 M, and (c) in SCW with new optimized inhibitor.

EDS Analysis

EDS analysis of samples after 48 h exposure to solution SCW without inhibitor and with inhibitor 1000 ppm molybdate and new optimized inhibitor is shown in Figure 11. The presence of C, Si, Mn, and Fe element can be observed in all figures. But the presence of Mo is only in Figures 11(b) and 11(c), indicating that molybdate oxide is formed on the surface of mild steel sample as stabilizer through its incorporation into the oxide film [18].

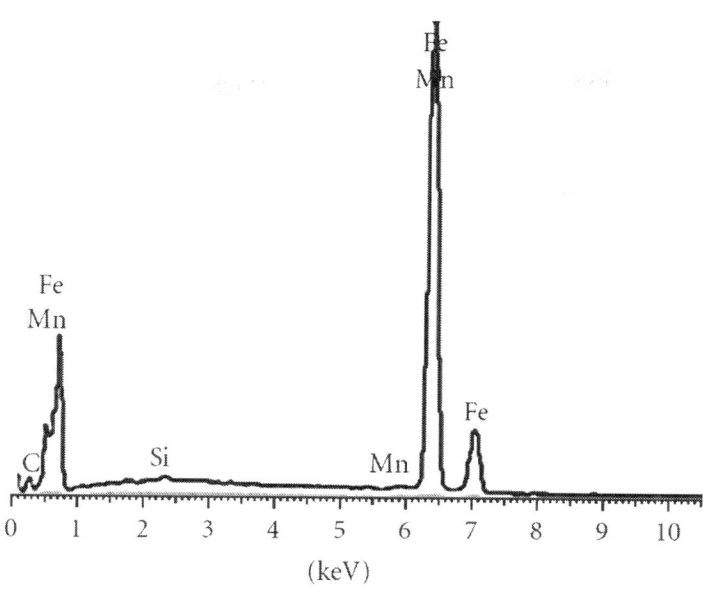

(a)

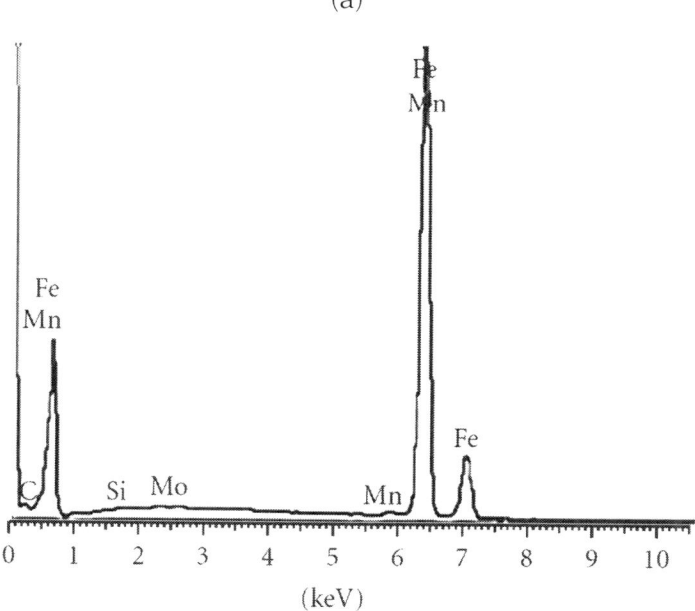

(b)

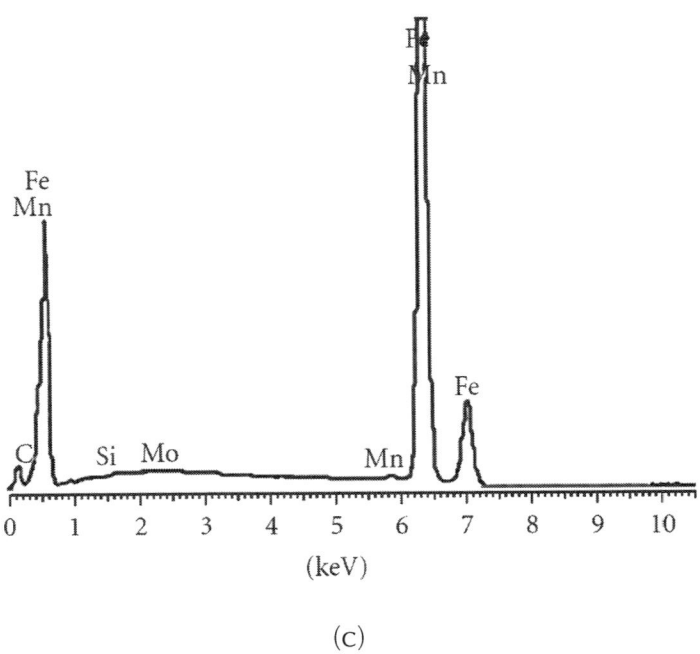

(c)

Figure 11: EDS analysis of mild steel (a) in SCW without inhibitor, (b) in SCW with inhibitor 1000 M, and (c) in SCW with new optimized inhibitor.

In addition, as shown in Figure 11(c), the carbon peak is much higher than (1000 ppm) molybdate alone, indicating that nitroethane ions are adsorbed on the surface besides molybdate ions to improve its inhibition efficiency synergically.

Finally, a glance at the electrochemical (polarization, AC impedance and weight loss method) and surface analysis (SEM photograph and EDS microanalysis) results indicates that new optimized inhibitor has relatively better efficiency than 1000 ppm molybdate alone at one ninth of the concentration range of molybdate, which is economically favorable.

CONCLUSIONS

- MoO_4^{2-} was proved to be a weak oxidizer, so another oxidizing agents such as NO_2^- and nitroethane beside cathodic inhibitor zinc were shown to be required to provide adequate protection for mild steel in stimulated cooling water.

- Results obtained from polarization curves showed that new optimized inhibitor with more than 98 percent inhibition efficiency noticeably lowers the corrosion rate of mild steel in stimulated cooling water.

- AC impedance curves showed that the combination of molybdate, nitrite, and nitroethane, beside cathodic inhibitor zinc in optimized range can be a suitable replacement for high concentration of 1000 ppm molybdate.

- The results of weight loss measurements obtained from coupon testing specimens exhibit a good correlation with data obtained from the polarization and AC impedance curves.

- The presence of molybdate as stabilizer of oxide film is observed through the EDS analyzer.

- SEM images indicate that new optimized inhibitor could form a relatively steady, compact, and uniform film on the surface of mild steel.

REFERENCES

1. C. M. Mustafa and J. P. G. Farr, "A potentiodynamic study of the corrosion inhibition of mild steel in realistic situation by molybdate and organic compounds containing –COOH and/or –OH groups,"Indian Journal of Technology, vol. 30, p. 424, 1992.

2. O. Lahodny-Sarc, F. Kapor, and R. Halle, "Corrosion inhibition of carbon steel in chloride solutions by blends of calcium gluconate and sodium benzoate," Materials and Corrosion, vol. 51, no. 3, pp. 147–151, 2000. ·

3.   Y. J. Qian and S. Turgoose, "Inhibition by zinc-molybdate mixtures of corrosion of mild steel," British Corrosion Journal, vol. 22, no. 4, pp. 268–271, 1987.

4.   M. J. Pryor and M. Cohen, "The inhibition of the corrosion of iron by some anodic inhibitors," Journal of the Electrochemical Society, vol. 100, pp. 203–215, 1953.

5.   E. A. Lizlovs, "Molybdates as corrosion inhibitors in the presence of chlorides," Corrosion, vol. 32, no. 7, pp. 263–266, 1976. ·

6.   P. A. Burda, "Molybdates as chromate replacement for closed cooling. Water systems in nuclear industry," Corrosion, vol. 92, p. 118, 1992.

7.   M. A. Stranick, "Corrosion inhibition of metals by molybdate. Part I. mild steel," Corrosion, vol. 40, no. 6, pp. 296–302, 1984. ·

8.   A. J. Bentley, L. G. Earwaker, J. P. G. Farr, and A. M. Seeney, "A technique for the in situ elemental analysis of electrode surfaces," Surface Technology, vol. 23, no. 1, pp. 99–103, 1984. ·

9.   J. P. G. Farr and M. Saremi, "Molybdate in aqueous corrosion inhibition I: effects of molybdate on the potentiodynamic behaviour of steel and some other metals," Surface Technology, vol. 19, no. 2, pp. 137–144, 1983.

10.   C. M. Mustafa and S. M. S. I. Dulal, "Molybdate and nitrite as corrosion inhibitors for copper-coupled steel in simulated cooling water," Corrosion, vol. 52, no. 1, pp. 16–22, 1996.

11.   D. B. Alexander and A. A. Moccari, "Evaluation of corrosion inhibitors for component cooling water systems," Corrosion, vol. 49, no. 11, pp. 921–928, 1993.

12.   V. S. Sastri, Corrosion Inhibitors. Principles and Applications, John Wiley & Sons, Toronto, Canada, 1998.

13.   J. Jefferies and B. Bucher, "New look at molybdate," Materials Performance, vol. 31, no. 5, pp. 50–53, 1992. ·

14.   S. Karim, C. M. Mustafa, M. D. Assaduzzman, and M. Islam,

"Effect of nitrate ion on corrosion inhibition of mild steel in simulated cooling water," Chemical Engineering Research Bulletin, vol. 14, pp. 87–91, 2010.

15. F. Mansfeld, M. W. Keding, and S. Tsai, "Recording and analysis of AC impedance data for corrosion studies-experimental approach and results," Corrosion, vol. 38, p. 301, 1982.

16. W. J. Lorenz and F. Mansfeld, "Determination of corrosion rates by electrochemical DC and AC methods," Corrosion Science, vol. 21, no. 9-10, pp. 647–672, 1981.

17. ASTM, "Standard practice for laboratory immersion corrosion testing of metal," ASTM International G 31-72, West Conshohocken, Pa, USA, 1990.

18. S. M. A. Shibli and V. A. Kumary, "Inhibitive effect of calcium gluconate and sodium molybdate on carbon steel," Anti-Corrosion Methods and Materials, vol. 51, no. 4, pp. 277–281, 2004. ·

Interface Control of Atomic Layer Deposited Oxide Coatings by Filtered Cathodic Arc Deposited Sublayers for Improved Corrosion Protection

Emma Härkönen[a], Sanna Tervakangas[b], Jukka Kolehmainen[b], Belén Díaz[c], Jolanta Światowska[c], Vincent Maurice[c], Antoine Seyeux[c], Philippe Marcus[c], Martin Fenker[d], Lajos Tóth[e], György Radnóczi[e], and Mikko Ritala[a]

[a]Laboratory of Inorganic Chemistry, University of Helsinki, FIN-00014 Helsinki, Finland

[b]DIARC-Technology Inc., Espoo, Finland

[c]Laboratoire de Physico-Chimie des Surfaces, CNRS (UMR 7075) – Chimie ParisTech (ENSCP), F-75005 Paris, France

[d]FEM Research Institute, Precious Metals and Metals Chemistry,

D-73525 Schwäbisch Gmünd, Germany
eResearch Centre for Natural Sciences HAS, (MTA TKK), Budapest, Hungary

ABSTRACT

Sublayers grown with filtered cathodic arc deposition (FCAD) were added under atomic layer deposited (ALD) oxide coatings for interface control and improved corrosion protection of low alloy steel. The FCAD sublayer was either Ta:O or Cr:O–Ta:O nanolaminate, and the ALD layer was Al_2O_3–Ta_2O_5 nanolaminate, $Al_xTa_yO_z$ mixture or graded mixture. The total thicknesses of the FCAD/ALD duplex coatings were between 65 and 120 nm. Thorough analysis of the coatings was conducted to gain insight into the influence of the FCAD sublayer on the overall coating performance. Similar characteristics as with single FCAD and ALD coatings on steel were found in the morphology and composition of the duplex coatings. However, the FCAD process allowed better control of the interface with the steel by reducing the native oxide and preventing its regrowth during the initial stages of the ALD process. Residual hydrocarbon impurities were buried in the interface between the FCAD layer and steel. This enabled growth of ALD layers with improved electrochemical sealing properties, inhibiting the development of localized corrosion by pitting during immersion in acidic NaCl and enhancing durability in neutral salt spray testing.

GRAPHICAL ABSTRACT

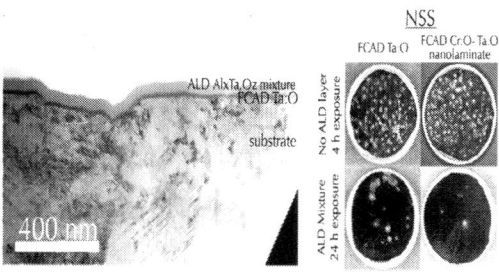

INTRODUCTION

Corrosion protection of engineering metals and alloys with atomic layer deposited (ALD) oxide coatings has gained increasing attention during the last years. Protective layers have been deposited on stainless steel [1], [2], [3], [4], [5] and [6], steel [7], [8], [9], [10], [11], [12], [13] and [14], aluminium alloy [7], magnesium alloy [15], magnesium–lithium alloy [16], copper [17] and [18] and silver [19]. Because ALD is based on alternating precursor pulses separated by inert gas purging, film growth occurs only on surfaces [20] and [21]. This leads to high conformality and uniformity even on challenging surface morphologies. Therefore ALD grown protective oxide layers offer significant advantages over ceramic coatings deposited on e.g. steel by many other methods [22], [23], [24], [25], [26], [27] and [28]. Morphological heterogeneities, which pose a problem for physical vapour deposition (PVD) [23] and [24], do not influence the quality of the coating and even complicated 3D objects can be coated conformally with ALD. Moreover, because complete burial of surface heterogeneities is not necessary, the ALD coatings can be considerably thinner than PVD coatings. Post-deposition annealing treatments, which are usually necessary with solution deposition techniques like sol–gel and can lead to crack formation [26], are not needed in ALD. Also intrinsic defect formation during the coating process, which is typical for instance for plasma electrolytic oxidation (PEO) [28], is not an issue for ALD. Thus very low pinhole and other defect densities can be accomplished. Furthermore, combining two or more materials into nanolaminates or mixtures allows easy modification of the composition and architecture of the coatings for the best combination of properties.

The ALD thin film materials that have been considered as corrosion protection coatings on steel are Al_2O_3, TiO_2 and Ta_2O_5 [1], [2], [3], [4], [5], [6], [7], [8], [9], [10], [11], [12], [13] and [14]. The best sealing properties in electrochemical measurements were achieved with Al_2O_3 [1], [4], [7], [8] and [9]. With 50 nm thin films deposited at 250 °C a three orders of magnitude decrease in passive current density of a stainless steel was obtained [4]. Similarly, on a low alloy steel a 50 nm Al_2O_3 coating deposited at 160 °C decreased the corrosion current density by two orders of magnitude [8]. More moderate sealing properties were observed with the TiO_2 and Ta_2O_5 coatings [1], [2], [3], [4], [5], [6], [10], [11] and [12]. Unfortunately, Al_2O_3 was observed to dissolve from a steel surface at a rate of 7 ± 1 nm per hour even in neutral NaCl solutions [9]. The dissolution was attributed to cathodic reduction of dissolved oxygen at the bottom of pinholes resulting in a local increase of pH. The stability of the coatings could be improved by combining the insulating properties of Al_2O_3 with the chemical stability of TiO_2 or Ta_2O_5 [1], [3], [4], [5], [6], [10], [12] and [13]. The best long-term corrosion protection properties were achieved with Al_2O_3–TiO_2 nanolaminate, Al_2O_3–Ta_2O_5 nanolaminate and AlxTayOz mixture coatings.

The ALD film growth begins with chemical reactions between the precursors and the substrate surface [20] and [21]. Therefore the chemical species on the surface have an effect on the quality of the film deposited on top: a lack of appropriate surface species for the film growth can induce nucleation delays and poor adhesion, impurities and weakly attached particles can induce defect formation, and hydrocarbon impurities can lead to poor adhesion and sealing properties [7], [13], [14], [29] and [30]. The compositionally and morphologically heterogeneous industrial metal alloys do not offer the best starting surfaces for ALD film growth. Impurities and loose particles are hard to avoid and often the surfaces contain some type of a hydrocarbon layer. It has been observed that coatings deposited with thermal ALD on steel substrates that have been cleaned only by degreasing in organic solvents have problems with adhesion [7], [13] and [14]. Pre-treatment with H_2–Ar plasma was found to have a beneficial effect on the coating-steel interface, and thus both adhesion and electrochemical barrier properties of the coatings could be improved [14]. Additionally, the stability of the coatings has been shown to improve with decreasing substrate roughness [13]. As ALD growth occurs uniformly and conformally over all surfaces, the improvement was most likely due to a reduction in the number of weakly attached particles, which upon detaching can create a pinhole defect in the coating. However, even better properties can be expected when the ALD films are grown on clean, well-defined surfaces with appropriate starting points for the film growth.

Filtered cathodic arc deposition (FCAD) is a PVD technique [31] and [32]. It is based on a low-voltage, high-current plasma discharge between two metallic electrodes. The plasma discharge brings forth an arc current composed of high-energy ions and electrons. A part of the ion flux is directed to a substrate after magnetically filtering away macroparticles formed in the plasma. The film deposition occurs through bombardment of the substrate with the high-energy ion flux. This leads to films with excellent adhesion, high density and hardness. The process can also involve an *in situ* pre-cleaning step that removes impurities like hydrocarbons and oxide layers from the substrate surface. FCAD coatings are widely used as hard protective coatings for reducing mechanical wear [33] and [34]. The characteristics of FCAD make it an ideal candidate to be combined with ALD for resolving the challenging aspects of the solely ALD-based protective coatings on metallic substrates.

In this study, we have combined the advantageous properties of FCAD with ALD films by preparing thin (≤120 nm) FCAD/ALD duplex coatings for corrosion protection of steel. Careful attention was given to the effect of the FCAD sublayers on the morphology, composition, electrochemical properties, stability and long-term corrosion protection properties of the ALD coatings. Two FCAD sublayers, 10 nm Ta:O and 50 nm Cr:O–Ta:O nanolaminate, were employed [11] and [35]. The top ALD layers were 50 nm Al_2O_3–Ta_2O_5 nanolaminate and AlxTayOz mixtures with either homogenous or graded composition as selected based on previously published results [10] and [12].

EXPERIMENTAL

Low alloy steel (AISI 52100, DIN 100Cr6) hardened and tempered (at 180 °C) to 805 HV hardness was used as a substrate material. The composition of the steel was (in wt.%) C (0.95–1.1), Cr (1.5), Ni (max. 0.30), Mn (0.25–0.45), Cu (max. 0.30), Si (0.15–0.35), P (max. 0.030), S (max. 0.025) and Fe (balance). The substrate surfaces were tumble polished, ground by planar grinding, lapped in a water based diamond suspension (6 μm) and brushed.

The FCAD coating process was carried out in a DIARC-Technology Inc. coating equipment. The deposition sequence was the same as presented in previous publications [11] and [35]. Before coating the samples were wiped with acetone, ultrasonicated in isopropanol for 5 min, rinsed with isopropanol and blow-dried with compressed air. Then they were etched *in situ* in the FCAD chamber with 350 eV Ar ions at 0.5 mA cm^{-2} current density for 30 min. The metal oxide coatings, Ta:O and Cr:O, were produced from Cr and Ta plasma in presence of low partial pressure of oxygen. The deposition temperature was below 100 °C.

Prior to ALD the samples were once more wiped with acetone, ultrasonicated in acetone and isopropanol for 5 min, rinsed with ethanol and blow-dried with compressed air. Further purification of the surface was done by H_2–Ar plasma at 160 °C in a Beneq TFS-200 ALD reactor according to methodology detailed in a previous publication [14]. The plasma was generated by a capacitively coupled 13.56 MHz rf power source. The reactor was operated in a remote plasma configuration, i.e. the plasma was separated from the substrates by a metal grid. The plasma gases Ar (>99.999%) and H_2 (>99.999%) were purified on site with Aeronex Gatekeeper and Entergris Gatekeeper purifiers. The gas flows were maintained constant at 130 and 15 sccm for Ar and H_2. The pre-treatment was conducted by ALD type pulsing to avoid excessive temperature increase during the plasma treatment. The plasma was turned on for 5 s and off for 10 s. The cycle was repeated 360 times to reach the desired 30 min pre-treatment time. The plasma power was 170 W. The pre-treatments were conducted *ex situ*, i.e. after the pre-treatment the reactor was cooled to 100 °C, opened to normal laboratory air and the samples were moved into the reactor used for the ALD deposition as fast as possible. The approximate air exposure time was 2–3 min.

ALD coatings were grown in a Picosun SUNALE R-150 reactor at 160 °C. The ALD process details were the same as in previous publications [10] and [12]. The precursors were trimethyl aluminium $(Al(CH_3)_3$, TMA, Chemtura AXION® PA 1300), tantalum pentaethoxide $(Ta(OC_2H_5)_5$, SAFC Hitech™) and ultrapure water $(H_2O$, resistivity > 18 MΩ cm). TMA and H_2O were evaporated at room temperature and Ta $(OC_2H_5)_5$at 140 °C. The pulse lengths were 0.1 s in Al_2O_3 and 0.4 s in Ta_2O_5 deposition sequence. The purge was always 5 s. The growth rate of Al_2O_3 was 0.09 nm cycle^{-1} and Ta_2O_5 0.04 nm cycle^{-1}. The number of cycles was chosen so that the nominal 50 nm coating thickness was reached.

The coatings prepared, their coding and nominal thicknesses are presented in Table 1. The FCAD coatings are coded by a letter F and the ALD coatings by a letter A. Thereafter the FCAD and ALD coatings are numbered from 1 to 3. Duplex coatings bear the coding of both the FCAD and the ALD layers.

Table 1: Coding and nominal thicknesses of studied coatings on steel

Code	FCAD layer	ALD layer
F1	10 nm Ta:O	–
F2	50 nm Ta:O	–

F3	2 × [10 + 10] nm Cr:O–Ta:O nanolaminate + 10 nm Cr:O	–
F1–A1	10 nm Ta:O	2 × [12.5 + 12.5] nm Al_2O_3–Ta_2O_5 nanolaminate
F1–A2	10 nm Ta:O	50 nm AlxTayOz mixture
F1–A3	10 nm Ta:O	50 nm AlxTayOz graded mixture
F3–A1	2 × [10 + 10] nm Cr:O–Ta:O nanolaminate + 10 nm Cr:O	2 × [12.5 + 12.5] nm Al_2O_3–Ta_2O_5 nanolaminate
F3–A2	2 × [10 + 10] nm Cr:O–Ta:O nanolaminate + 10 nm Cr:O	50 nm AlxTayOz mixture
F3–A3	2 × [10 + 10] nm Cr:O–Ta:O nanolaminate + 10 nm Cr:O	50 nm AlxTayOz graded mixture

The FCAD and ALD coating thicknesses were measured from silicon samples coated simultaneously with the steel substrates. The measurements were conducted with a Dektak 3ST profilometer and x-ray reflectance (XRR, Bruker AXS D8 Advance diffractometer) for the FCAD and ALD thin films, respectively. The XRR curves were modelled with Leptos 7.05.

The pristine morphology of the coatings was studied with field emission scanning electron microscopy (FESEM, Hitachi S-4800) and transmission electron microscopy (TEM, Philips CM20). FESEM imaging was used to study the surface morphology of the steel before and after coating with FCAD and ALD. TEM was used for cross sectional imaging of the samples. Prior to TEM analysis, the samples were thinned by standard mechanical grinding and ion bombardment techniques: the samples were cut, embedded into a Ti-holder, mechanically ground and polished, and finally milled with 10 keV Ar^+ ions. The final step of the ion milling was carried out with 3 keV to minimize the damage to the thinned samples.

The composition of the coatings and the coating-substrate interface was studied with time-of-flight secondary ion mass spectrometry (ToF-SIMS). A ToF-SIMS 5 spectrometer (IonToF) was employed. The measurements were done with a pulsed 25 keV Bi^+ primary ion source delivering 0.8 pA of analysis current over a 100 × 100 μm^2 area. The depth profiling was done by sputtering with a 2 keV Cs^+ beam giving a target current of 82 nA over a 400 ×

400 μm^2 area. Negative ion profiles were used as they are more sensitive to fragments from oxide matrixes. The operation pressure was 10^{-9} mbar. Ion-Spec software was used for the data acquisition and post-processing.

Polarization (linear sweep voltammetry, LSV) measurements were used for evaluating the electrochemical properties of the coatings. The measurements were conducted with an AUTOLAB PGSTAT30 potentiostat/galvanostat at room temperature in a 0.2 M NaCl solution (Analar Normpur analytical reagent VWR® BDH Prolabo®) at pH 7. The electrolyte solution was bubbled with Ar for 30 min prior to starting and throughout the measurements. A three-electrode system with platinum as the counter electrode and standard calomel electrode (SCE) as the reference was used. The electrochemical measurements were always started with 30 min open circuit potential (OCP) measurement to ensure the stability of the system. Polarization was measured from −0.9 V up until the anodic current density exceeded 10 μA cm^{-2} with a scan rate of 1 mV s^{-1}. The exposed sample area was limited to 0.44 cm^2 with a Viton o-ring. Corrosion current densities and corrosion potentials were obtained by Tafel analysis [36] and [37].

The stability of the coatings was evaluated by immersion in a 0.2 M NaCl solution at pH 2 (0.01 M HCl) (NaCl and HCl 37% Analar Normpur analytical reagent VWR® BDH Prolabo®). The total immersion time was 6 h. During immersion electrochemical impedance spectroscopy (EIS) measurements were conducted at regular intervals with the AUTOLAB PGSTAT30 potentiostat/galvanostat. The EIS measurements were done at OCP with the exciting signal amplitude set to 10 mV. The frequency range was from 10^{-2}–10^5 Hz. The experimental impedance spectra were modelled with the ZSimp-Win software based on the minimization of the ^{2} function, defined as the sum of the squares of the differences between the measured and the calculated data. After the stability testing the ToF-SIMS depth profile analysis was repeated to gain insight into the compositional changes of the samples. The depth profiles were measured as detailed above. Corrosion durability was studied with a neutral salt spray (NSS) test according to the standard DIN 50021 (ISO 9227) with the exception that the samples were removed from the chamber at regular intervals, rinsed with deionised water and photographed. During NSS the temperature, NaCl concentration and pH were kept constant at 35 ± 2 °C, 50 ± 5 g l^{-1} and 6.5–7.2, respectively. The extent of corrosion in per cents after 2, 4, 24 and 48 h of exposure was quantified according to the Renault standard D17 1058J. A grid consisting of 4×4 mm squares was placed on the sample and the number of squares containing corrosion spots was considered against the total number of squares. The whole square was considered corroded if even one corrosion spot could be found in it. At the edges of the circular samples only tiles filled with over 50% by the sample were considered. Rust grading for the samples was then given according to percentages in standard DIN 51802 (Table 2).

Table 2: Definition of rust grades according to standard DIN 51802

Rust grade	Description of the rust figure	Area of corrosion/%
0	No corrosion	0
1	Maximum 3 corrosion spots covering less than 1 mm²	Not defined
2	Slight corrosion	<1
3	Moderate corrosion	1–5
4	Heavy corrosion	5–10
5	Very heavy corrosion	>10

RESULTS AND DISCUSSION

Coating Morphology and Composition

In the FESEM images the bare substrate surface appeared heterogeneous (Fig. 1a). Scratches, holes and particles could be observed. The 50 nm ALD mixture coating (A2) or nanolaminate (not shown) alone covered the surface conformally slightly blurring the scratches (Fig. 1b), as was observed previously [10] and [12] and expected based on the conformal ALD growth mechanism [20] and [21]. The 10 nm FCAD Ta:O coating (F1) followed the surface of the substrate closely (Fig. 1c), but also made some of the scratches and holes more pronounced. A smoother surface was observed with the 50 nm FCAD Cr:O–Ta:O nanolaminate (F3) coating (Fig. 1d) that buried some surface irregularities. The ALD top layers in the duplex coatings smoothened the surface further by conformal coverage (Fig. 1e and f). No defects could be observed in any of the studied coatings (Fig. 1).

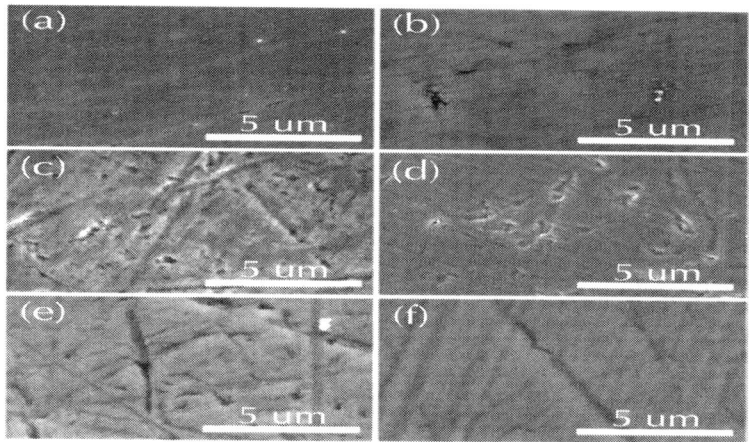

Figure 1: FESEM images of (a) bare, (b) ALD AlxTayOz mixture (A2), (c) FCAD Ta:O (F1), (d) FCAD Cr:O–Ta:O nanolaminate (F3), (e) FCAD Ta:O + ALD AlxTayOz mixture (F1–A2) and (f) FCAD Cr:O–Ta:O nanolaminate + ALD AlxTayOz mixture (F3–A2) coated steel.

The FCAD and ALD layers were clearly visible and distinguishable in the cross sectional TEM images (Fig. 2). The interfaces were sharp and both layers followed the surface conformally. The duplex coatings appeared to be well adhered to the substrate and the layers to each other, confirming previous results for single coatings [11], [14] and [35]. All the FCAD and ALD layers were amorphous, as expected [38]. Direct electron diffraction patterns could not be obtained because the coatings were very thin, but when looking at the bright and dark field contrast only amorphous structures were seen. No pinholes or other defects could be observed with the local TEM cross sectional analysis. The FCAD layers were slightly thicker than their nominal values suggested. The "10 nm Ta:O" layer (F1) was ~15 nm thick and the "50 nm Cr:O–Ta:O nanolaminate" layer (F3) ~70 nm. The ALD layers were close to the nominal 50 nm thickness.

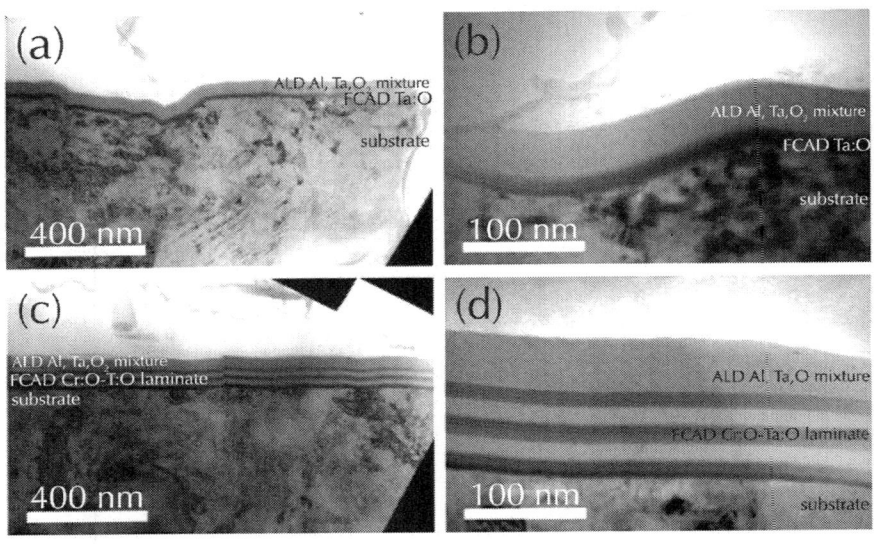

Figure 2: TEM cross sectional images of duplex FCAD/ALD coated steel: (a and b) FCAD Ta:O + ALD AlxTayOz mixture (F1–A2) and (c and d) FCAD Cr:O–Ta:O nanolaminate + ALD AlxTayOz mixture (F3–A2).

ToF-SIMS depth profiles of the FCAD Ta:O combined with the ALD nanolaminate (F1–A1), mixture (F1–A2) and graded mixture (F1–A3) layers are presented in Fig. 3. The depth profiles had similar general features as single FCAD and ALD coatings on steel [8], [9], [10], [11], [12], [13], [14] and [35]. The coating and interface regions could be easily distinguished. The interface starting point was determined from the start of an increase of Fe$^-$ and Cr$^-$ ion profiles. Different from the single ALD coatings and similar to the single FCAD coatings on steel, no peak in the FeO$_2^-$, CrO$_2^-$, Fe$^-$

and Cr⁻ ion profiles could be observed at the coating-substrate interface [8], [9], [10], [11], [12], [13], [14] and [35]. The Ar ion etch process prior to FCAD appeared to have completely etched away any native oxide on the steel surface, as discussed previously [11] and [35]. The FCAD coating also suppressed formation of a new interface layer by oxidation of steel during exposure air and in the initial stages of the ALD process. Native oxide suppression and inhibition of oxide regrowth were confirmed by the TEM data (Fig. 2) that did not show the approximately 10 nm thick interface layer observed between the single ALD coatings and substrate [8], [9], [10], [11], [12],[13] and [14]. Instead TaC⁻ and C⁻ ToF-SIMS signals appeared to peak at the FCAD coating-steel interface. This can be attributed to a formation of a Ta/Ta–C interlayer due to a reaction of the carbonaceous impurities remaining on the substrate with the first tantalum ions arriving in the beginning of the FCAD process [11] and [35].

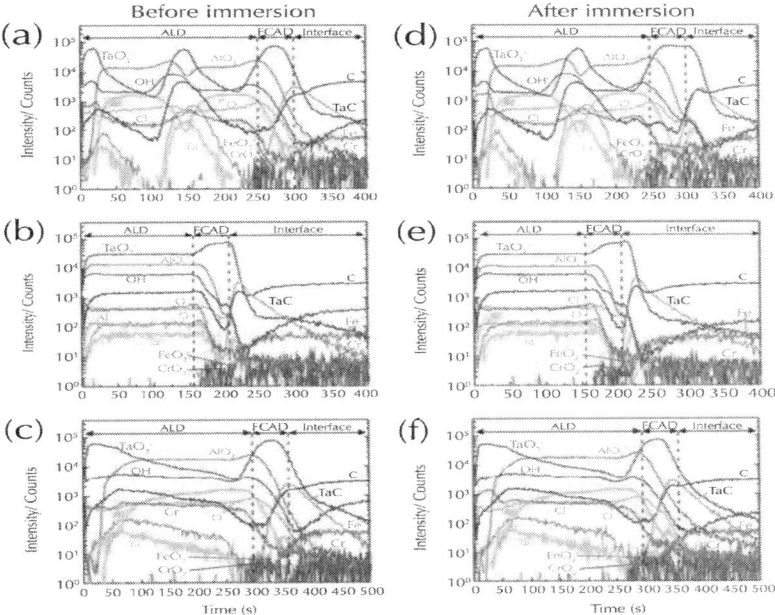

Figure 3: ToF-SIMS depth profiles of duplex FCAD/ALD coated steel before and after immersion in 0.2 M NaCl solution at pH 2: (a and d) Ta:O

+ Al_2O_3–Ta_2O_5 nanolaminate (F1–A1), (b and e) Ta:O + AlxTayOz mixture (F1–A2) and (c and f) Ta:O + AlxTayOz graded mixture (F1–A3).

The coating region could be further divided into separate ALD and FCAD layers (Fig. 3). The end of the ALD top layer and start of the FCAD sublayer was taken as the point where TaO_2^- signal started to increase or AlO_2^- signal started to decrease near the interface with the substrate. The sputtering times for reaching the FCAD layer differed for the three samples. Rather than being an indication of different thicknesses of the ALD layers, this was likely due to matrix effects encountered when measuring depth profiles of films with different compositions [39]. The FCAD layers and their interfaces to steel seemed to remain unaffected by the ALD process on top of them, and only low OH^-, Cl^- and C^- impurity signals could be seen (Fig. 3). No signal from the substrate species could be seen in the entire duplex coating region indicating that the coatings were pinhole free at least in the resolution of the ToF-SIMS equipment used.

In the depth profile of the FCAD Ta:O and ALD nanolaminate (F1–A1) duplex coating the different layers of the ALD nanolaminate could be easily distinguished (Fig. 3a). Clear peaking of Al^-, AlO_2^-, Ta^- and TaO_2^- ion intensities could be observed in the corresponding Al_2O_3 and Ta_2O_5 layers. The peaks of the different layers appeared to overlap. Rather than implicating that the layers were mixed, this was probably due to the roughness of the substrate [4], [10] and [13]. The C^- impurity signal peaked with the Ta_2O_5 layers, and the OH^- signal at the interfaces between the Al_2O_3 and Ta_2O_5 layers. This was observed also previously with the single ALD nanolaminates on steel, and mirrors the higher amount of impurities in the ALD Ta_2O_5 compared to Al_2O_3 [10] and [13]. Peaking of Cl^- could be seen in the Ta_2O_5 layers. In the ALD Al_2O_3 coatings on steel the chlorine contamination has been assigned to a ≤0.01 wt% dimethylaluminium chloride (DMACl) impurity content of the aluminium precursor TMA [12]. The exact origin of the chlorine contamination in the ALD Ta_2O_5 coatings on steel has not been cleared [12]. However, the total Cl^- contamination in both Al_2O_3 and Ta_2O_5 has been shown to be below the detection limit 0.5 at.%

of X-ray photoelectron spectroscopy (XPS) [8] and [11]. The depth profile of the ALD top layer in the FCAD Ta:O and ALD mixture (F1–A2) duplex coating appeared very similar to a single ALD mixture coating on steel (Fig. 3b) [12] and [13]. The Al^-, AlO_2^-, Ta^- and TaO_2^- signals were constant throughout the ALD layer thickness, showing homogeneous in-depth composition, and C^-, OH^- and Cl^- impurities were observed.

The depth profile of the FCAD Ta:O and ALD graded mixture (F1–A3) duplex coating on steel is presented in Fig. 3c. In the ALD top layer a fast decrease of Ta^- and TaO_2^- could be observed indicating that the composition changed monotonously through the coating from Ta_2O_5 to Al_2O_3 as designed. The Al^- and AlO_2^- signals changed also along the depth profile, but the change was not as fast as for the Ta_2O_5 species. The differences in the slopes were most probably due to matrix effects [39]. The C^- impurity signal decreased with the Ta_2O_5 species along the coating thickness confirming the higher carbon contamination of ALD Ta_2O_5 compared to Al_2O_3 [10]. The OH^- and Cl^- impurity signals were approximately constant throughout the thickness.

Barrier Properties of the Coatings

The $i–E$ polarization curves of uncoated, single FCAD and duplex FCAD/ALD coated steel in 0.2 M NaCl solutions at pH 7 are presented in Fig. 4. The polarization curve of the uncoated steel indicated that the anodic reaction was activation controlled, and the cathodic reaction was under diffusion control near the corrosion potential and under activation control at the most negative potentials [36]. The active anodic behaviour was expected as the steel contains only low amounts of Cr (1.5 wt.%). The diffusion controlled cathodic reaction was the reduction of dissolved oxygen, concentration of which was very low in the Ar bubbled electrolyte solution. The activation controlled cathodic reaction was hydrogen reduction. The corrosion current density of the uncoated steel was determined by Tafel analysis [36] and [37] to be 4.6×10^{-7} A cm^{-2} (Table 3).

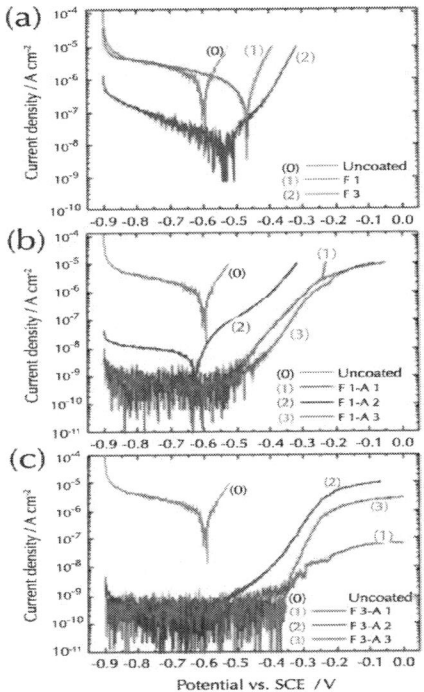

Figure 4: Polarization results of single FCAD and duplex FCAD/ALD coated steel: (a) FCAD Ta:O (F1) and Cr:O–Ta:O nanolaminate (F3), (b) FCAD Ta:O + ALD Al_2O_3–Ta_2O_5 nanolaminate (F1–A1)/ AlxTayOz mixture (F1–A2)/ AlxTayOz graded mixture (F1–A3) and (c) FCAD Cr:O–Ta:O nanolaminate + ALD Al_2O_3–Ta_2O_5 nanolaminate (F3–A1)/ AlxTayOz mixture (F3–A2)/ AlxTayOz graded mixture (F3–A3). The polarization curve of the uncoated steel is included in all images for reference.

Table 3: Tafel analysis results from polarization measurements on bare, single FCAD and duplex FCAD/ALD coated steel

Sample	Corrosion current densi-ty/×10^{-9} A cm^{-2}	Porosity/%
Uncoated steel	460	100
F1	250	55

F3	5.3	1.1
F1–A1	0.35	0.08
F1–A2	9.4	2.0
F1–A3	0.23	0.05
F3–A1	0.21	0.05
F3–A2	0.19	0.04
F3–A3	0.32	0.07

The polarization curves of the single FCAD coated samples had similar characteristics as the uncoated steel (Fig. 4a). With both coatings (F1 and F3) the anodic reaction appeared to have shifted towards positive potentials indicating some ennoblement of the steel-coating system. As discussed previously [11] and [35], and confirmed by the present ToF-SIMS data, it is likely that the removal of the native oxide present on the uncoated alloy by pre-etching in the FCAD process promoted corrosion resistance of the reactive uncoated surface. The thicker FCAD Cr:O–Ta:O nanolaminate (F3) coating had an additional benefit of significantly decreasing the access of oxygen to the steel surface and thus reducing the corrosion current density by two orders of magnitude compared to the uncoated steel (Table 3).

The FCAD/ALD duplex coatings all had better protective properties than the single FCAD coatings (Fig. 4b and c). Similar behaviour as for the uncoated steel was observed for the duplex coatings. However, due to the extremely low current densities observed for most samples in the cathodic potential range, the reaction mechanisms could not be determined certainly. The current was at or below the detection limit of the equipment used for the measurements. The ALD mixture layers on the FCAD Ta:O and Cr:O–Ta:O nanolaminate layers (F1–A2 and F3–A2, respectively) appeared to slightly reduce the corrosion potential of the uncoated steel and to create a semi-passive range in the anodic branch. A similar response was observed with ALD Ta_2O_5 and sufficiently Ta_2O_5-rich mixtures on steel [11] and [12]. The behaviour was assigned to a growth of an unusually chromium-rich, considering the steel composition, corrosion product layer at the steel-coating interface during exposure of the samples to air after plasma pre-treatment or in the beginning of the ALD film growth. This phenomenon was less apparent with the ALD mixture on the thicker FCAD Cr:O–Ta:O nanolaminate layer (F3–A2), which indicates that in the F1–A2 sample the thin FCAD Ta:O layer was not sealing sufficiently to prevent interface oxidation. However, the FCAD/ALD mixture duplex coatings showed also slightly inferior polarization behaviour compared to single ALD mixture on steel [12] and [13]. The poorer sealing properties of the ALD mixtures on the FCAD sublayers may be due to a detrimental interaction of the tantalum precursor with the FCAD layers. In particular, the oxygen deficient FCAD Ta:O layers [11] may promote decomposition of $Ta(OC_2H_5)_5$ leading to defects. Indeed, all the other FCAD/ALD duplex coatings (F1–A1, F1–A3, F3–A1 and F3–A3) where the ALD coating started with Al_2O_3 on the FCAD sublayer had better sealing properties than the corresponding single ALD coatings [12]. The difference to the duplex coatings with the ALD mixture top layers might be the prevention of $Ta(OC_2H_5)_5$ interacting with the oxygen deficient FCAD layers by the ALD Al_2O_3 starting layer, or the more sealing nature of ALD Al_2O_3 compared to Ta_2O_5 inhibiting more efficiently the oxidation of the steel surface through defects in the FCAD layer. In duplex coatings with ALD nanolaminate or graded mixture the only improvement obtained by the thicker FCAD sublayer appeared to be moving the anodic reaction to higher potentials and thus extending the low current density range to approximately 100 mV higher potentials.

The corrosion current densities of the duplex coatings were determined by Tafel analysis on the anodic branch and fitting of a line on the cathodic branch [12] and [40]. The corrosion current densities were obtained from the intersection of the fitted lines. Compared to uncoated steel a decrease of more than three orders of magnitude in the corrosion current densities could be achieved with the best FCAD/ALD duplex coatings (Table 3).

To simplify the comparison of the coatings and their sealing properties, coating porosities (P) were calculated from the polarization results. The porosities represent the surface fraction of the substrate exposed to the surrounding environment through defects. The analysis was made based on corrosion current densities according to a procedure adapted from Tato et al. [41] (Equation (1)).

$$P = \frac{i_{corr}}{i^0_{corr}} \times 100\%$$

(1)

where i^0_{corr} represents the corrosion current density of the bare substrate, and i_{corr} the corrosion current density of the coated substrate under evaluation.

The calculated porosities are presented in Table 3. The FCAD Ta:O coating (F1) had an unrealistically large porosity of 55% considering that the coating appeared continuous in FESEM and TEM (Fig. 1 and Fig. 2). The evaluation of porosities based on corrosion current densities assumes that the coating is inert and all current originates from the electrochemical behaviour of steel exposed through defects in the coating. However, the FCAD Ta:O coatings have been observed to be significantly oxygen deficient [11] and [35]. Thus the high porosity obtained for the continuous coating can be partially explained by a weak electrochemical activity of the coating. Porosities of the same order of magnitude were detected also previously for thin 10 nm FCAD Ta:O coatings on steel [35]. The FCAD Cr:O–Ta:O nanolaminate (F3) had a significantly lower porosity of 1.1%, owing to the higher thickness and possibly to the less oxygen deficient Cr:O layers [35].

All the FCAD/ALD duplex coatings had lower porosities than the corresponding single FCAD coating (Table 3). The ALD mixture on FCAD Ta:O (F1–A2) and Cr:O–Ta:O nanolaminate (F3–A2) coatings had porosities of 2.0 and 0.04%. As already discussed above, this implies a less effective or similar sealing as was observed with single ALD mixture on steel (0.04%) [12], possibly due to a detrimental reaction of the $Ta(OC_2H_5)_5$ ALD precursor with the oxygen deficient FCAD Ta:O [11]. The other duplex coatings, ALD nanolaminate on FCAD Ta:O (F1–A1) and Cr:O–Ta:O nanolaminate (F3–A1), and ALD graded mixture on FCAD Ta:O (F1–A3) and Cr:O–Ta:O nanolaminate (F3–A3), had porosities very close to each other (0.06 ± 0.02%). Significant improvement from the porosities of single ALD nanolaminate (0.7%) and graded mixture (0.2%) on steel [12] could be achieved indicating better sealing with the duplex coatings, however with a total thickness exceeding that of the single ALD counterparts. The thicker Cr:O–Ta:O nanolaminate sublayers (F3) appeared to offer little improvement compared to the thinner Ta:O sublayers (F1) under the ALD nanolaminate and graded mixture layers even though there was a significant difference between the porosities of the single FCAD layers. This implies that the 10 nm Ta:O layer was sufficient in homogenizing the interface for good ALD growth, and that the sealing properties were mainly obtained from the ALD layers.

Coating Stability and Performance during Immersion in Acidic Nacl

ToF-SIMS depth profiles of FCAD Ta:O and ALD duplex coatings on steel after a 6-h immersion in 0.2 M NaCl solution at pH 2 are presented in Fig. 3. Stabilities of single FCAD Ta:O and ALD Al_2O_3, Ta_2O_5, Al_2O_3–Ta_2O_5 nanolaminate and $Al_xTa_yO_z$ mixture coatings on steel in the same conditions have been studied [9], [11] and [13]. FCAD Ta:O coatings were found to be stable throughout the immersion in NaCl at pH 2 [11]. Immersion of the ALD Al_2O_3 in 0.2 M NaCl solution at pH 7 showed clear dissolution of the coating: in the ToF-SIMS analysis the sputtering time needed to reach the interface layer decreased linearly with the immersion time [9]. Chloride ions penetrated through the coating and accumulated at the interface. A simultaneous decrease of the hydroxide ion intensity suggested that the chloride ions replaced hydroxyl impurities in the coating. In contrast, single ALD Ta_2O_5 coatings were found stable in the 0.2 M NaCl solution at pH 7 and pH 2 [11]. However, similar penetration of chloride ions through the coating, as observed with Al_2O_3, was also observed with Ta_2O_5 coatings, and significant localized pitting corrosion occurred due to a less efficient sealing as compared to ALD Al_2O_3. With the AlxTayOz mixture and Al_2O_3–Ta_2O_5 nanolaminate coatings on steel in 0.2 M NaCl solution at pH 2 no general coating dissolution occurred [13]. However, in the ToF-SIMS depth profile of the AlxTayOz mixture a minor net increase of the interfacial oxide signals in the coating region was observed indicating some local coating breakdown. Also some chloride accumulation at the interface was shown. The Al_2O_3–Ta_2O_5 nanolaminate appeared to be more stable: no general or local breakdown was observed, but chloride ions were observed to penetrate into the outermost Ta_2O_5 layers of the coating.

The depth profiles of the FCAD/ALD duplex coatings were nearly identical before and after immersion (Fig. 3). The sputtering times needed to reach the FCAD and interface regions were the same indicating that no general dissolution of the coatings occurred. The same compositional features and the same intensities could be observed. The only difference was an increase of the Cl^- intensity in the outermost parts of the ALD Ta_2O_5 layers (Fig. 5), as was seen also with the ALD Al_2O_3–Ta_2O_5 nanolaminate alone[13]. This behaviour could be seen in the ALD nanolaminate (F1–A1) and graded mixture (F1–A3) topped coatings. In the ALD mixture topped coating the Cl^- intensity was the same before and after immersion. It is well known that ALD Ta_2O_5 is more porous than ALD Al_2O_3 [4], [10], [11], [12] and [42], thus enabling Cl^-penetration only in the outermost Ta_2O_5 layers of the nanolaminate and graded mixture as observed in the present study. The decrease of the porosity when reaching an Al_2O_3 layer (nanolaminate) or a sufficiently Al_2O_3 rich layer (graded mixture) hindered further penetration.

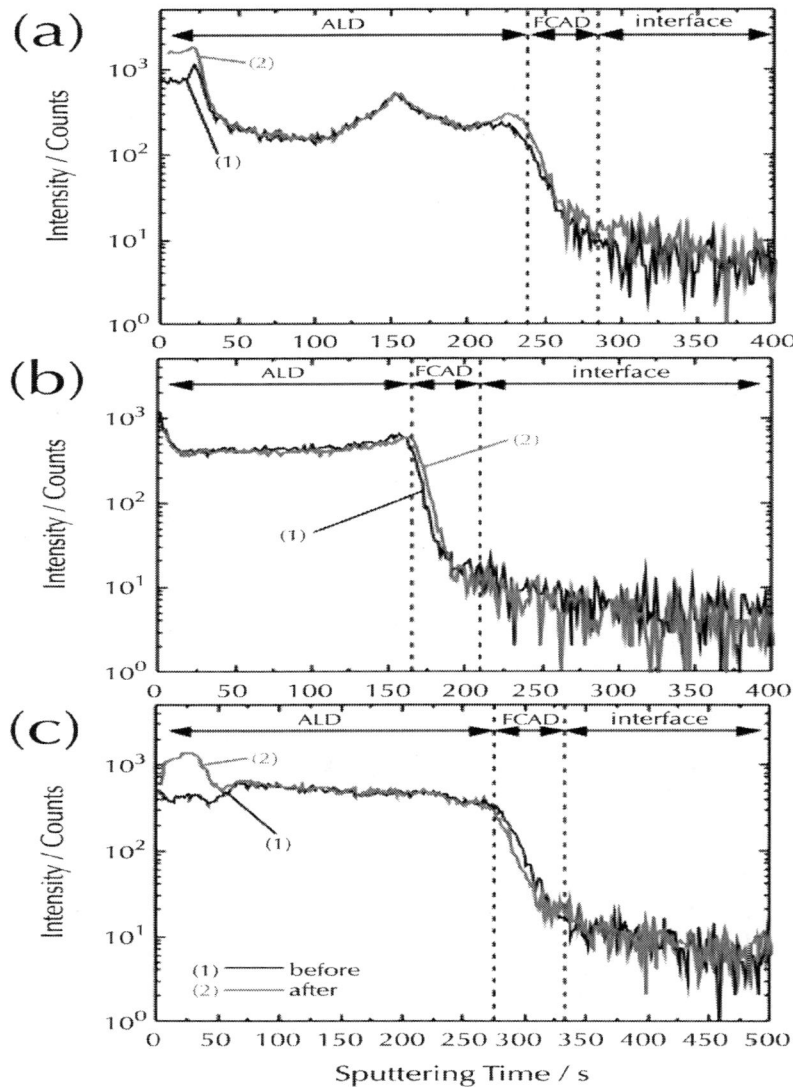

Figure 5: ToF-SIMS Cl⁻ depth profiles of duplex FDAD/ALD coated steel before and after immersion in 0.2 M NaCl solution at pH 2: (a) FCAD Ta:O + ALD Al_2O_3–Ta_2O_5 nanolaminate (F1–A1), (b) FCAD Ta:O + ALD AlxTayOz mixture (F1–A2) and (c) FCAD Ta:O + ALD AlxTayOz graded mixture (F1–A3).

The general appearances of the EIS Bode plots during immersion in 0.2 M NaCl solution at pH 2 were similar for all the duplex FCAD Ta:O and ALD coated samples (Fig. 6). At the high frequencies the spectra were constant throughout the immersion. At the middle frequencies the impedances appeared constant, but the phase angles were slightly modified at the frequencies between 10^{-1} and 10^2 Hz. This is different from what was observed for single ALD Al_2O_3 coatings in neutral NaCl solutions where modifications appeared in the whole frequency range [9]. With Al_2O_3 coatings the modifications were assigned to variation of the coating thickness and to redox reactions occurring on the substrate surface at the bottom of pinholes. Another phenomenon must be occurring on the duplex samples as no variation of the coating thickness was observed with ToF-SIMS and the differences in phase angle modification imply a process with a different time constant. On steel coated with single ALD Ta_2O_5, Al_2O_3–Ta_2O_5 nanolaminate and AlxTayOz mixture layers and on anodized aluminium alloys and aluminium alloys passivized in $CeCl_3$ solutions, changes in the frequency range from 10^{-1}–10^2 Hz have been assigned to pitting corrosion [11],[13], [43] and [44]. At low frequencies a decrease of global impedance was observed, indicating a decrease of protective properties of the coatings during the immersion. The only clear difference between the spectra of the three FCAD/ALD duplex coatings was the greater initial resistance to corrosion, but a larger modification later after two hours of immersion, of the ALD nanolaminate topped sample compared to the other two duplex coatings. For the ALD mixture and graded mixture the decrease of global impedance was smaller and approximately constant throughout the immersion.

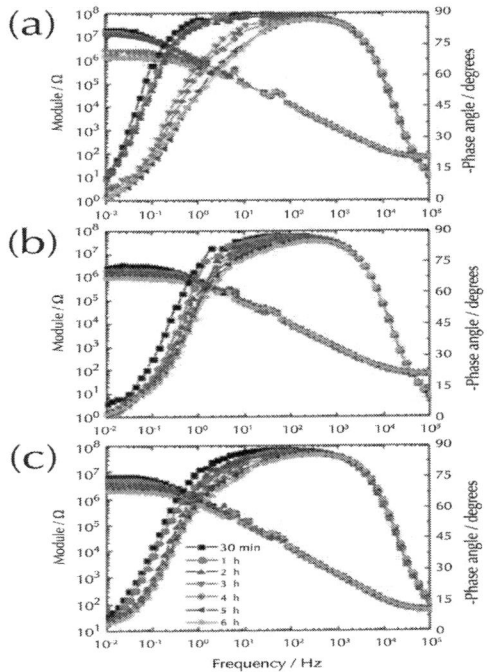

Figure 6: EIS Bode plots of steel coated with duplex FCAD/ALD coatings during immersion in 0.2 M NaCl solution at pH 2: (a) FCAD Ta:O + ALD Al$_2$O$_3$–Ta$_2$O$_5$ nanolaminate (F1–A1), (b) FCAD Ta:O + ALD AlxTayOz mixture (F1–A2) and (c) FCAD Ta:O + ALD AlxTayOz graded mixture (F1–A3).

The Nyquist plots gave some further information on the corrosion phenomena on the sample surfaces (Fig. 7). In addition to the high frequency capacitive semicircle, mostly responsible for the phenomena observed in the Bode plots, an inductive loop could also be seen at the low frequencies. During immersion this inductive loop appeared to start changing into a capacitive semicircle. This was also observed with ALD Ta$_2$O$_5$, Al$_2$O$_3$–Ta$_2$O$_5$ nanolaminate and Al$_x$Ta$_y$O$_z$ mixture coatings on steel in acidic NaCl[11] and [13]. In literature the inductive to capacitive transformation is usually assigned to increasing corrosion product accumulation on iron or steel surfaces in acidic solutions [45] and [46]. With the ALD Ta$_2$O$_5$

and AlxTayOz mixture coated steel the accumulation of FeO_2^- and CrO_2^- at the bottom of pinholes at the coating-substrate interface was also observed with ToF-SIMS [11] and [13]. However, no such increase could be observed with ToF-SIMS for the duplex coatings (Fig. 3). It is probable that the FCAD layer suppressed the oxidation of the interface so much that it could not be detected with ToF-SIMS.

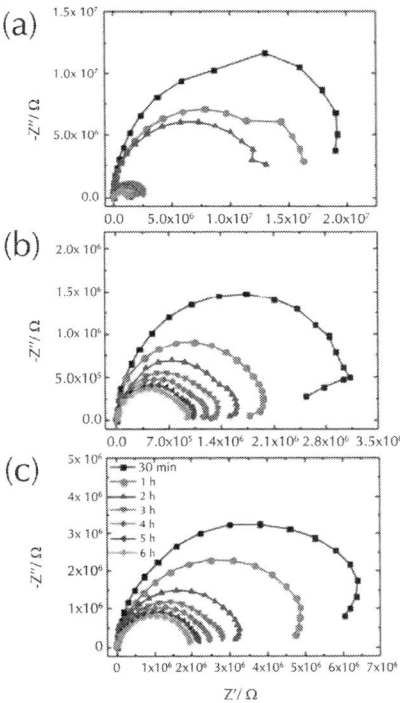

Figure 7: EIS Nyquist plots of steel coated with duplex FCAD/ALD coatings during immersion in 0.2 M NaCl solution at pH 2: (a) FCAD Ta:O + ALD Al_2O_3–Ta_2O_5 nanolaminate (F1–A1), (b) FCAD Ta:O + ALD AlxTayOz mixture (F1–A2) and (c) FCAD Ta:O + ALD AlxTayOz graded mixture (F1–A3).

The equivalent circuit used for the data modelling is presented in Fig. 8. It is a modification of the circuit previously discussed by Bonnel et al. [47] and Díaz et al. [9], [11] and [13]. The coated surface is characterized by a coating capacitance, C_{coat}. The uncoated surface fraction, i.e. steel surface exposed through coating pinholes, is represented by charge transfer resistance, R_{ct}, in parallel with a series of resistance of pitting corrosion, R_{pit} and double layer capacitance, C_{dl}. R_{pit} represents defects that evolve during the immersion while R_{ct} represents the continuously corroding steel surface. The solution resistance, R_e, is in series with all these components. This circuit was chosen mainly on the base of previous EIS results on immersion of single FCAD or ALD coated steel in NaCl where two time constants were very clearly distinguishable from each other [9], [11] and [13]. The inductive loop observed in the Nyquist plots in the beginning of immersion (Fig. 7) was not considered in the fitting as further discussion on it is beyond the aim of the present paper. All the capacitances are presented by constant phase elements (CPE), with which the un-ideal capacitive behaviour caused by surface heterogeneity can be considered [48]. The impedance of a CPE is represented by Equation (2) [49].

$$Z_{CPE} = \frac{1}{Q(jw)^n}$$

(3)

where Q and n can be obtained directly from the fitting. The value of Q is independent of frequency, and the factor n represents the CPE power, which is related in Nyquist plots to the angle of depression of the semicircle beneath the x-axis by $(1 - n) \times 90°$. The real capacitances can be evaluated using the Brug approach (Equation (3)) [50].

$$C = Q^{1/n}\left(\frac{1}{R_{ct}} + \frac{1}{R_e}\right)^{\frac{n-1}{n}}$$

(4)

where the contribution of R_{ct} is negligible due to $R_{ct} \gg R_e$. The fitted values after 0.5, 1, 3 and 6 h of immersion are shown in Table 4. The capacitances calculated using Equation (3) are presented instead of CPEs.

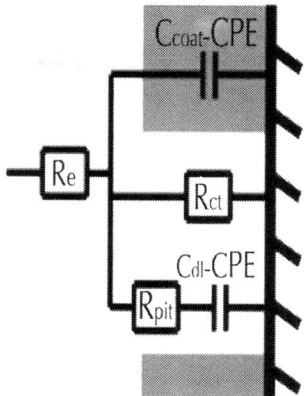

Figure 8: Equivalent circuit used for EIS data modelling. The symbols R_e, R_{ct} and R_{pit} represent resistances of the electrolyte solution, the charge transfer and the pitting, respectively. The symbols CPE_{coat} and CPE_{dl} represent the constant phase elements of the coating and the double layer at the steel surface.

Table 4: EIS fitting results on duplex FCAD/ALD coated steel during immersion in 0.2 M NaCl solution at pH 2

	Time	R_e/Ω	C_{coat}/F	n_1	R_{ct}/Ω	C_{dl}/F	R_{pit}/Ω	n_2
F1–A1	30 min	68	1.4e-7	1.0	2.3e7	1.5e-8	2.0e6	0.80
	1 h	67	1.4e-7	1.0	1.5e7	2.4e-8	7.1e5	0.74
	3 h	65	1.3e-7	1.0	2.9e6	1.2e-7	6.0e5	0.72
	6 h	65	1.3e-7	1.0	1.4e6	2.9e-7	2.6e5	0.70
F1–A2	30 min	71	1.6e-7	0.98	3.4e6	6.3e-8	8.7e5	0.64
	1 h	71	1.6e-7	0.98	2.1e6	6.2e-8	6.5e5	0.60
	3 h	70	1.6e-7	0.98	1.4e6	1.3e-7	2.9e5	0.60
	6 h	70	1.6e-7	0.98	9.3e5	1.9e-7	2.3e5	0.73
F1–A3	30 min	69	1.2e-7	0.98	7.8e6	4.9e-8	7.6e6	0.71
	1 h	68	1.2e-7	0.98	5.4e6	7.9e-8	1.4e6	0.76
	3 h	65	1.2e-7	0.98	2.8e6	9.2e-8	2.8e5	0.73
	6 h	64	1.2e-7	0.98	2e6	1.5e-7	2.9e5	0.79

The corrosion rate can be considered based on the charge transfer resistances, R_{ct} (Table 4). All duplex coatings had larger R_{ct} values than the single FCAD or ALD coatings and significantly higher values than the uncoated steel [9], [11], [13] and [35]. This indicates that the duplex coatings were more efficient in inhibiting the corrosion of steel than the single FCAD or ALD layers. In the beginning of immersion the duplex coatings could be arranged in order based on their protective properties: F1–A1 > F1–A3 > F1–A2. The ALD nanolaminate topped sample showed significantly better protective properties than the mixtures in the beginning of the immersion. In the end of the immersion the order had changed to F1–A3 > F1–A1 > F1–A2, and the differences in the corrosion resistances of the samples had diminished. A similar trend was observed with single ALD nanolaminate and mixture coatings on steel [13]. The end-of-immersion order is in line with that of the polarization results (Table 3). Since the polarization measurements are more destructive than EIS, the better correspondence of the polarization results with the EIS results after 6 h rather than at the beginning of immersion can be accounted for.

The pitting resistances, R_{pit}, correspond to local modifications in the pinholes exposing the steel surface, i.e. pitting corrosion [43] and [44]. The R_{pit} values of the duplex coatings both at the beginning and end of immersion could be arranged in order F1–A3 > F1–A1 > F1–A2 (Table 4), which corresponds with the polarization results and charge transfer resistances at the end of immersion. The R_{pit} values of all the duplex coatings were very close to each other at the end of immersion indicating that pitting corrosion was similar.

Both resistances R_{ct} and R_{pit} were observed to be higher and to decrease less during the immersion for the FCAD/ALD duplex-coated samples than for the single ALD nanolaminate and mixture coatings [13]. The improvement was most pronounced in the 10^{-1}–10^2 Hz frequency range, i.e. in the pitting resistance. This indicates that the FCAD sublayer significantly inhibited the pitting process. It was observed above (Fig. 3) that the FCAD layers suppressed the formation of an interfacial oxide layer and buried the hydrocarbon contamination, and it is known that ALD film growth is sensitive to the substrate surface state [13], [14],[29] and [30]. Thus the improved resistance to pitting was probably due to improved initial ALD growth on a more homogeneous and less contaminated surface.

The coating capacitances, C_{coat}, remained constant throughout the immersion (Table 4). As previously discussed [9], [11], [13] and [51], this is an indication that the coatings were stable and no general dissolution occurred confirming the ToF-SIMS data. Additionally the CPE powers were very close to unity showing that the coating surfaces were homogenous and smooth [48]. The capacitances of the duplex coatings were quite close to each other. The values increased in order F1–A3 < F1–A1 < F1–A2. If a parallel plate capacitor structure is assumed, the capacitances, C, can be evaluated from Equation (4) [52].

$$C = \frac{\varepsilon_0 \varepsilon_r A}{d}$$

(4)

where ε_0 represents vacuum dielectric constant (8.85×10^{-14} F cm^{-1}), ε_r dielectric constant of the coating material, A area and d thickness. Amorphous ALD Al_2O_3 and Ta_2O_5 deposited at 250 °C are known to have approximate dielectric constants of 8 and 21 [53] and [54]. The content of Ta_2O_5 is also smaller in the graded mixture than in the mixture ALD coating [12]. Therefore if the area and thickness of the coatings were the same, the capacitance values corresponded well with the general composition of the coatings.

The double layer capacitances, C_{dl}, of all the coatings increased slightly during the immersion (Table 4). If a parallel plate capacitor is again assumed (Equation (4)), the result implies an increase of surface area in steel exposed to solution. Because the R_{pit} is connected in series with the C_{dl} in the equivalent circuit, and the R_{pit} values decreased during immersion, an increase of pit density and/or size was implied. The CPE powers for the double layer capacitances were between 0.8 and 0.6 with no clear trends detected. CPE powers between 0.8 and 0.6 are usually taken to represent general corrosion [46]. The small fluctuation probably implied some changes in the exposed surface morphology at the bottom of the coating pinholes.

Corrosion Protection Properties

The NSS durability results on single FCAD and duplex FCAD/ALD coated steel are presented in Fig. 9 andTable 5. The single FCAD coated samples were completely covered by corrosion spots and had a rust grade of 5 after two hours of testing. After four hours the number of corrosion spots appeared almost the same, but their size had increased. In an earlier publication it was observed that on steel coated with single ALD nanolaminate, mixture and graded mixture the first corrosion spots appeared after two, four and two hours, respectively [12]. However, the extent of corrosion was quite low still after 24 h of testing with all the single ALD coatings.

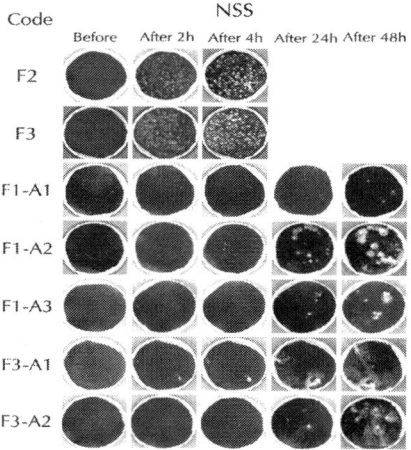

Figure 9: NSS results on steel protected with single FCAD and duplex FCAD/ALD coatings: FCAD Ta:O (F2), FCAD Cr:O–Ta:O nanolaminate (F3), FCAD Ta:O + ALD nanolaminate (F1–A1), FCAD Ta:O + ALD mixture (F1–A2), FCAD Ta:O + ALD graded mixture (F1–A3), FCAD Cr:O–Ta:O nanolaminate + ALD nanolaminate (F3–A1) and FCAD Cr:O–Ta:O nanolaminate + ALD mixture (F3–A2).

Table 5: Rust grades of single FCAD and duplex FCAD/ALD coated steel during NSS testing

Code	NSS			
	After 2 h	After 4 h	After 24 h	After 48 h
F2	5	5		
F3	5	5		
F1–A1	0	0	1	4
F1–A2	0	0	5	5
F1–A3	0	0	5	5
F3–A1	5	5	5	5
F3–A2	1	1	5	5

All the samples with the FCAD/ALD duplex coatings had better corrosion durability than the samples with single FCAD or ALD coatings (Fig. 9 and Table 5) [12]. On samples with the FCAD Ta:O combined with the ALD nanolaminate, mixture or graded mixture the first corrosion spots appeared after 24 h of testing. Even after 48 h the samples had only scattered corrosion spots and large areas on the sample surfaces remained corrosion-free. The best protective properties were achieved with the ALD nanolaminate top coating (F1–A1). On samples with the FCAD Cr:O–Ta:O nanolaminate under the ALD nanolaminate or mixture the corrosion started already after two hours of testing. However, the extent of corrosion was extremely low. With the naked eye only a few corrosion spots could be observed on samples coated with both types of ALD coatings. After 48 h of testing the samples appeared similar to the ALD coatings combined with the thinner FCAD Ta:O sublayer, and the sample surfaces were still mostly corrosion free. Overall similar results were achieved with all the duplex coatings after 48 h of immersion. These results support the conclusion made from the porosities that the thicker FCAD sublayer did not offer significant improvement to the duplex coatings. The main benefits of the FCAD sublayers were obtained already with the thin Ta:O layers.

Significant improvement to the single FCAD or ALD coating durability was achieved with the duplex coatings (Fig. 9) [12]. However, the first corrosion spots could be observed after 24 h of NSS testing even in the best cases (Fig. 9 and Table 5). Many industrial applications require >100 h NSS durability rendering even the FCAD/ALD duplex coatings unsuitable. Thus the long-term durability of the coating systems should be further developed. Additionally, the ultra thin (≤120 nm) ceramic layers cannot withstand mechanical wear. Sealing of thick PVD coatings on steel by ALD has been previously suggested [55], [56],[57] and [58]. Another possibility would be to first seal a surface with ALD or FCAD/ALD duplex coating, and thereafter grow a thick PVD layer on top.

CONCLUSIONS

In this work the beneficial effect of adding thin (≤70 nm) FCAD sublayer beneath an ALD oxide coating for improved corrosion protection of low alloy steel has been demonstrated. Combined FESEM, TEM, ToF-SIMS, LSV and EIS analysis was conducted, and coating durability was assessed with NSS testing. The coatings were found to be conformal and uniform, and the adhesion between the FCAD and ALD layers and substrate appeared sufficient. The different layers in the coatings could be easily distinguished both by cross-sectional and depth profile analysis. The bulk regions of the FCAD and ALD layers were similar in composition to the corresponding single coatings on steel. The FCAD process was observed to efficiently remove and suppress the formation of a layer of iron/chromium oxide at the coating-substrate interface and to bury residual carbonaceous impurities of the substrate surface in the FCAD-substrate interface. As a result, both the electrochemical sealing properties and NSS durability of the duplex coatings were improved compared to single FCAD and ALD coatings. Immersion tests of the duplex-coated steel in acidic NaCl solutions showed that the initial charge transfer and pitting resistances were higher than for the single layer coated samples and the decrease of the resistances during the immersion was slower. Additionally, the duplex coatings were stable and compositionally almost unaffected by the immersion.

The beneficial influence of the FCAD sublayer appeared to arise from a better control of the interface. The homogenous FCAD oxide without significant carbonaceous contamination enabled a more ideal start for the ALD film growth. The coatings thus had fewer weak points, which are likely sites for initiation and development of localized corrosion by pitting in aggressive environments. Hence the FCAD sublayer inhibited the nucleation and/or growth of pits at the least protected sites, thus slowing down the development of localized corrosion and improving the protection properties of the coatings.

ACKNOWLEDGMENTS

The research leading to these results has received funding from the European Community's Seventh Framework Program (FP7/2007-2013) under grant agreement n° CP-FP 213996-1. Academy of Finland (Finnish Centre of Excellence in Atomic Layer Deposition) is also thanked for support. Region Ile-de-France is acknowledged for partial support for the ToF-SIMS equipment.

REFERENCES

1.   R. Matero, M. Ritala, M. Leskel a, T. Salo, J. Aromaa, O. Forsen, J. Phys. IV Fr.ance 9 (1999). Pr8-493ePr8-499.

2.   C.X. Shan, X. Hou, K.-L. Choy, Surf. Coat. Technol. 202 (2008) 2399e2402.

3.   E. Marin, A. Lanzutti, F. Andreatta, M. Lekka, L. Guzman, L. Fedrizzi, Corros. Rev. 29 (2011) 191e208.

4.   B. Díaz, J. Swiatowska, V. Maurice, A. Seyeux, B. Normand, E. H arkonen, M. Ritala, P. Marcus, Electrochim. Acta 56 (2011) 10516e10523.

5.   E. Marin, A. Lanzutti, L. Guzman, L. Fedrizzi, J. Coat. Technol. Res. 8 (2011) 655e659.

6.   E. Marin, L. Guzman, A. Lanzutti, W. Ensinger, L. Fedrizzi, Thin Solid Films 522 (2012) 283e288.

7.   S.E. Potts, L. Schmalz, M. Fenker, B. Díaz, J. Swiatowska, V. Maurice, A. Seyeux, P. Marcus, G. Radnoczi, L. T oth, W.M.M. Kessels, J. Electrochem. Soc. 158 (2011) C132eC138.

8.   B. Díaz, E. H arkonen, J. Swiatowska, V. Maurice, A. Seyeux, P. Marcus, M. Ritala, Corros. Sci. 53 (2011) 2168e2175.

9.   B. Díaz, E. H arkonen, J. Swiatowska, V. Maurice, A. Seyeux, M. Ritala, P. Marcus, Electrochim. Acta 56 (2011) 9609e9618.

10.   E. Hark onen, B. Díaz, J. Swiatowska, V. Maurice, A. Seyeux, M. Vehkam aki, T. Sajavaara, M. Fenker, P. Marcus, M. Ritala, J. Electrochem. Soc. 158 (2011) C369eC378.

11.   B. Díaz, J. Swiatowska, V. Maurice, A. Seyeux, E. H arkonen, M. Ritala,   S. Tervakangas, J. Kolehmainen, P. Marcus, Electrochim. Acta 90 (2013) 232e245.

12.   E. H arkonen, B. Díaz, J. Swiatowska, V. Maurice, A. Seyeux, M. Fenker, L. Toth, G. Radnoczi, P. Marcus, M. Ritala, Chem. Vapor Depos. 19 (2013) 194 e203.

13.   B. Díaz, E. Hark onen, J. Swiatowska, A. Seyeux, V. Maurice, M. Ritala, P. Marcus, Corros. Sci. 82 (2014) 208e217.

14.   E. Hark onen, S. Potts, W.M.M. Kessels, B. Díaz, A. Seyeux, J. Swiatowska, V. Maurice, P. Marcus, G. Radnoczi, L. T oth, M. Kariniemi, J. Niinist o, M. Ritala,   Thin Solid Films 534 (2013) 384e393.

15.   E. Marin, A. Lanzutti, L. Guzman, L. Fedrizzi, J. Coat. Technol. Res. 9 (2012) 347e355.

16.   P.C. Wang, Y.T. Shih, M.C. Lin, H.C. Lin, M.J. Chen, K.M. Lin, Thin Solid Films 518 (2010) 7501e7504.

17.   A.I. Abdulagatov, Y. Yan, J.R. Cooper, Y. Zhang, Z.M. Gibbs, A.S. Cavanagh, R.G. Yang, Y.C. Lee, S.M. George, ACS Appl. Mater. Interfaces 3 (2011) 4593e4601. Table 5 Rust

18.   M.L. Chang, T.C. Cheng, M.C. Lin, H.C. Lin, M.J. Chen, Appl.

Surf. Sci. 258 (2012) 10128e10134.

19. L. Paussa, L. Guzman, E. Marin, N. Isom aki, L. Fedrizzi, Surf. Coat. Technol. 206 (2011) 976e980.

20. M. Ritala, M. Leskela, Atomic layer deposition, in: H.S. Nalwa (Ed.), Handbook of Thin Film Materials, Academic Press, San Diego, 2001, pp. 103e158.

21. M. Ritala, J. Niinisto, Atomic layer deposition, in: A.C. Jones, M.L. Hitchman (Eds.), Chemical Vapour Deposition: Precursors, Processes and Applications, The Royal Society of Chemistry, Cambridge, 2009, pp. 158e206.

22. V.K.W. Grips, V.E. Selvi, H.C. Barshilia, K.S. Rajam, Electrochim. Acta 51 (2006) 3461e3468.

23. Y. Dianran, H. Jining, W. Jianjun, Q. Wanqi, M. Jing, Surf. Coat. Technol. 89 (1997) 191e195.

24. M. Cekada, P. Panjan, D. Kek-Merl, M. Panjan, G. Kapun, Vacuum 82 (2008) 252e256.

25. D. Wang, G.P. Bierwagen, Prog. Org. Coat. 64 (2009) 327e338.

26. G. Ruhi, O.P. Modi, A.S.K. Sinha, I.B. Singh, Corros. Sci. 50 (2008) 639e649.

27. W.-C. Gu, G.-H. Lv, H. Chen, G.-L. Chen, W.-R. Feng, G.-L. Zhang, S.-Z. Yang, J. Alloys Compd. 430 (2007) 308e312.

28. W. Gu, D. Shen, Y. Wang, G. Chen, W. Feng, G. Zhang, S. Fan, C. Liu, S. Yang, Appl. Surf. Sci. 252 (2006) 2927e2932.

29. Y. Zhang, D. Seghete, A. Abdulagatov, Z. Gibbs, A. Cavanagh, R. Yang, S. George, Y.-C. Lee, Surf. Coat. Technol. 205 (2011) 3334e3339.

30. M.D. Groner, J.W. Elam, F.H. Fabrequette, S.M. George, Thin Solid Films 413 (2002) 186e197.

31. I.G. Brown, Annu. Rev. Mater. Sci. 28 (1998) 243e269.

32. B.K. Tay, Z.W. Zhao, D.H.C. Chua, Mater. Sci. Eng. R 52 (2006) 1e48.

33. A. Anders, Vacuum 67 (2002) 673e686.

34. S. PalDey, S.C. Deevi, Mater. Sci. Eng. A 342 (2003) 58e79.

35. B. Díaz, J. Swiatowska, V. Maurice, M. Pisarek, A. Seyeux, S. Zanna, S. Tervakangas, J. Kolehmainen, P. Marcus, Surf. Coat. Technol. 206 (2012) 3903e3910.

36. L.J. Korb, Metals Handbook, ninth ed., vol. 13, ASM International, Ohio, 1987.

37. M. Stern, A.L. Geary, J. Electrochem. Soc. 104 (1957) 56e63.

38. V. Miikkulainen, M. Leskel a, M. Ritala, R. Puurunen, Appl. Phys. Rev. 113 (2013) 021301.

39. A.M. Belu, D.J. Graham, D.G. Castner, Biomaterials 24 (2003) 3635e3653.

40. R. Hausbrand, B. Bolando-Escudero, A. Dhont, J. Wielant, Corros. Sci. 61 (2012) 28e34.

41. W. Tato, D. Landolt, J. Electrochem. Soc. 145 (1998) 4173e4181.

42. K. Kukli, J. Ihanus, M. Ritala, M. Leskela, J. Electrochem. Soc. 144 (1997) 300e306.

43. F. Mansfeld, S. Lin, S. Kim, H. Shih, J. Electrochem. Soc. 137 (1990) 78e82.

44. H. Herrera-Hernandez, J.R. Vargas-Garcia, J.M. Hallen-Lopez, F. Mansfeld, Mater. Corros. 58 (2007) 825e832.

45. I. Ebelboin, M. Keddam, J.C. Lestrade, Faraday Discuss. Chem. Soc. 56 (1973) 264e275.

46. P. Li, T.C. Tan, J.Y. Lee, Corros. Sci. 38 (1996) 1935e1955.

47. A. Bonnel, J. Babosi, C. Deslouis, M. Duprat, M. Keddam, B. Tribollet, J. Electrochem. Soc. 130 (1983) 753e761.

48. J.-B. Jorcin, M.E. Orazem, N. Peb ere, B. Tribollet, Electrochim. Acta 51 (2006) 1473e1479.

49. E. Barsukov, J.R. Macdonald, Impedance Spectroscopy: Theory, Experiment and Applications, second ed., John Wiley & Sons Inc., Hoboken, New Jersey, 2000.

50. G.J. Brug, A.L.G. van den Eeden, M. Sluyters-Rebach, J.H. Sluyters, J. Electroanal. Chem. 176 (1984) 275e295.

51. C. Corfias, N. Peb ere, C. Lacabanne, Corros. Sci. 41 (1999) 1539 e1555.

52. W.D. Callister, Materials Science and Engineering: an Introduction, seventh ed., John Wiley & Sons, Inc., United States of America, 2007.

53. R. Matero, A. Rahtu, M. Ritala, M. Leskela, T. Sajavaara, Thin Solid Films 368 (2000) 1e7.

54. K. Kukli, M. Ritala, M. Leskela, J. Electrochem. Soc. 142 (1995) 1670 e1675.

55. C.X. Shan, X. Hou, K.-L. Choy, P. Choquet, Surf. Coat. Technol. 202 (2008) 2147e2151.

56. E. Marin, L. Guzman, A. Lanzutti, L. Fedrizzi, M. Saikkonen, Electrochem. Commun. 11 (2009) 2060e2063.

57. P.C. Wang, T.C. Cheng, H.C. Lin, M.J. Chen, K.M. Lin, M.T. Yeh, Appl. Surf. Sci. 270 (2013) 452e456.

58. E. Hark onen, I. Kolev, B. Díaz, J. Swiatowska, V. Maurice, A. Seyeux, P. Marcus, M. Fenker, L. Toth, G. Radnoczi, M. Vehkam aki, M. Ritala, ACS Appl. Mater. Interfaces 6 (2014) 1893e1901.

Chapter 8

Corrosion Protection Behavior of AZ31 Magnesium Alloy with Cathodic Electrophoretic Coating Pretreated by Silane

Jin Zhang and Chaoyun Wu

School of Materials Science and Engineering, University of Science and Technology Beijing, Beijing, 100083, PR China

ABSTRACT

As an alternative process to phosphate and chromate conversion coatings, silane pretreatment was used to improve the performance of cathodic electrophoretic coating (E-coat) on AZ31 Mg alloy in this study. The galvanic corrosion behavior of AZ31 Mg alloy with E-coat coupled with Q235 steel was investigated. Compared to bare Mg alloy and Mg alloy with conventional painting, the corrosion properties of the AZ31 Mg alloy pretreated with silane and subsequently E-coated were studied during salt solution immersion and salt spray testing. The surface morphologies of the Mg alloy were examined in detail after immersion in NaCl solution for different times using digital photography and scanning electron microscopy (SEM). The corrosion current density of the specimens was characterized by DC polarization tests. It was found that silane pretreatment of AZ31 Mg alloy followed by subsequent E-coat led to much better corrosion protection than that without silane treatment. The silane pretreatment and E-coat delayed the galvanic corrosion of Mg alloy coupled with 235 steel bolts.

INTRODUCTION

Mg is one of the lightest structured metals and its alloys have quite special properties which lead to many special applications. However Mg alloy is highly chemically reactive when exposed to humid environment or water and forms a loose oxide layer, which makes it very susceptible to corrosion. In most cases, atmospheric corrosion is usually uniform in industrial environments, whereas localized corrosion usually occurs under immersion conditions. Mg alloys are highly susceptible to galvanic corrosion when they are coupled with other metals as Mg has the lowest electrode potential compared to other alloys. As an anode, Mg alloys shows heavy localized corrosion.

To prevent or inhibit Mg alloy's corrosion, an appropriate surface pretreatment is required. Chromate conversion coatings have been used extensively as pretreatment methods for various Mg alloys to achieve good corrosion resistance and paint adhesion performance [1]. Due to their toxic and carcinogenic nature, however, chromate conversion coatings are being stringently limited and will soon be banned by environmental regulations for continuous use in coating processes [2]. Silane surface treatment is a promising environmentally friendly alternative pretreatment technique [3], [4], [5] and [6] for steel or aluminum alloy coated with painting primers, and also for Mg alloys [7]. Cathodic electrophoretic deposition (E-coat) is now being widely used in the automotive industry for the protection of steel and partly for aluminum. However, it is difficult to apply cathodic electrophoresis to bare Mg alloys for the following reasons: (1) Mg substrates are severely corroded by aqueous electrolytes; (2) Mg rapidly forms loose MgO and/or $Mg(OH)_2$ films, which inhibit successive electrodeposition and reduce the adhesion of the E-coat [8], [9], [10] and [11]; and (3) Mg alloys are easily dissolved in the acid solutions with pH < 7, which is the case for cathodic electrophoretic solutions. Consequently, a pretreatment must be carried out before E-coat is deposited onto Mg alloys.

Low-temperature plasma-deposited silane films followed by a subsequent E-coat has been used [12] for the corrosion protection of Mg. The plasma silane treatment is a "dry" process and can be performed in a vacuum reactor, which limits the size and shape of the Mg parts to be treated. To the best of our knowledge, the combination of a "wet" silane pretreatment [13] with subsequent E-coat on Mg alloys has never been reported except for the Chinese patent [14] up to now. In this study, therefore, the effect of a "wet" silane pretreatment followed by E-coating on the corrosion behavior of AZ31 Mg alloy has been investigated. The galvanic corrosion behavior of Mg alloy with silane/E-coat coupled with Q235 steel was additionally studied. It was anticipated that silane pretreatment would solve the problem of the fast dissolving of Mg alloy in the subsequent E-coat solutions [15]. Additionally, the silane film may increase the adhesion and corrosion resistance of E-coating on Mg substrates. In this paper our preliminary results on the corrosion protectiveness of the silane film, E-coat, and the silane/E-coat composite coating are reported. Our research aim is to solve the galvanic corrosion problem of Mg alloy coupled with Q235 steel bolts, which are often encountered in practical applications.

EXPERIMENTAL

Materials and Solutions

Specimens of AZ31 Mg alloy (nominal composition: 2.5–3.5% Al; 0.7–1.3% Zn; 0.2–1% Mn; <0.05% Si; <0.01% Cu; <0.001% Ni; <0.002% Fe) received from Southwest Aluminum Goods Manufacture Company (China) were used for the different treatments. For cleaning, Mg panels were first immersed in a commercial alkaline cleaner bath at 65 °C for 30 min, then in a pickling solution bath with 192 g/l CH_3COOH and 50 g/l $NaNO_3$ at ambient temperature for 1 min.

For the silane treatment, methanolic solutions of bis-[triethoxysilylpropyl]tetrasulfide silane with concentrations from 1% to 10% were prepared. All the silanic solutions were kept for more than 24 h before use.

For the E-coating, a commercial solution with epoxy resin from Institute of Southwest Engineering Technology was used.

Some pretreated samples were sent to a company, where the samples were painted like other commercial Mg alloy parts for comparison with the samples treated by E-coating or silane/E-coating of this study.

Silane Film and E-coat Preparation

For silanisation, the cleaned Mg panels were dipped in the silanic solution for 30 s, then dried by hot air to get rid of the excess liquid, and finally cured in an oven at 100 °C for 30 min to ensure condensation to form siloxanes.

For the cathodic electrophoretic deposition, stainless steel was used as anode and Mg alloy panel as cathode. The panels with and without silane film were placed individually into the container and set voltage at 200 V for 2 min. After E-coating, the panels were cleaned with DI-water and cured at 160 °C for half an hour.

Salt Spray Test

Two kinds of corrosion tests were performed: salt spray test was employed according to ASTM B117 specification. Aqueous salt solution with 5 wt.% NaCl and pH 7 was atomized in a salt spray chamber at 35 °C. The panels were placed in the chamber at an angle of 45°. Additionally, scribed test panels were immersed in an aqueous solution with 5% NaCl to evaluate the extent of corrosion performance and E-coat adhesion with or without silane pretreatment. The average width of coating delamination from the scribe after exposure was used to evaluate the extent of corrosion and adhesion of coating.

Galvanic Corrosion

Galvanic corrosion properties were studied by dipping the galvanic couple consisting of a Mg alloy panel and a Q235 mild steel bolt into 5 wt.% NaCl solution for 6 days. Three sets of specimens were tested, bare Mg alloy AZ31 (Mg), AZ31 Mg with E-coat (AZ31/E-coat), and AZ31 Mg with silane film and E-coat (AZ31/Silane/E-coat).

Corrosion Morphology and Microstructure Examination

The surface morphologies after the salt spray corrosion and galvanic corrosion tests were recorded with a Digital Camera and DSM-III macroscope. The microstructure of the observed corrosion pits was examined by a JSM-6460LV scanning electronic microscope.

DC Polarization Tests

DC polarization tests were carried out with AZ31 Mg alloy panels in a neutral 3.5 wt.% NaCl solution. Electrochemical measurements were performed using a Model 273A potentiostat. The working electrode was an E-coated Mg panel with a working area of 1 cm × 1 cm. A saturated calomel electrode (SCE) and a platinum plate were used as reference electrode and auxiliary electrode, respectively. The test was performed with a sweep rate of 1 mV/s and a sweep voltage of ±250 mV (with respect to open potential).

RESULTS AND DISCUSSION

Corrosion Resistance of the Silane Film

A colorless and transparent thin film was formed on the surface of AZ31 Mg alloy panels after the silane pretreatment. As shown in Fig. 1, the silane film was very uniform. With micro-arc anodization or phosphate coating [16] and [17], it was shown that the film had no pores. The SEM micrographs also show the underlying grain boundaries which were formed on the AZ31 Mg surface during acidic pickling before the silane treatment.

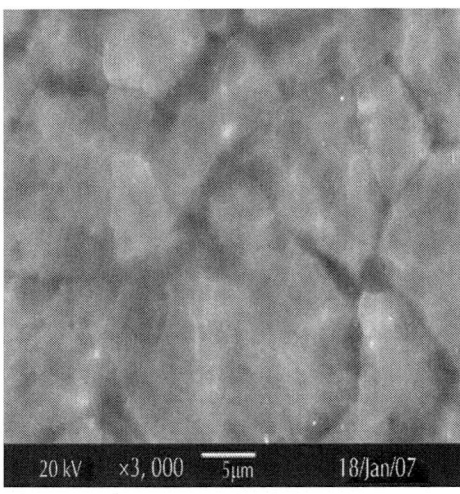

Figure 1: Film formed on AZ31B Mg alloy surface in 8% BTESPT silane solution for 50 s at room temperature.

The silanes formed Si–O–Si and Si–O–Mg bonds on the surface, as shown by FTIR spectroscopy[18] and [19]. The formation of a crosslinked silane film is attributed to the condensation/ crosslinking of silanol groups (SiOH), which are initially formed by the hydrolysis of ester groups ($-SiOC_2H_5$) in the dilute water/ alcohol silane solution [20]. The Si–O–Mg bonds lead to strong adhesion between the AZ31 Mg substrate and the silane film, as already reported in [18].

The silane film had good corrosion resistance and could also enhance the adhesion of subsequent primer coatings to Mg alloy substrates [21]. It can be seen from the polarization curve in Fig. 2 that bare AZ31 Mg had a slightly higher corrosion potential than silane treated AZ31 Mg alloy, the silane film decreased the corrosion currents density by more than one order of magnitude. The cathodic shift of the corrosion potential and the decrease of the cathodic current density indicate that cathodic inhibition likely occurred after silane treatment as reported in literature for these types of conversion films [22] and [23].

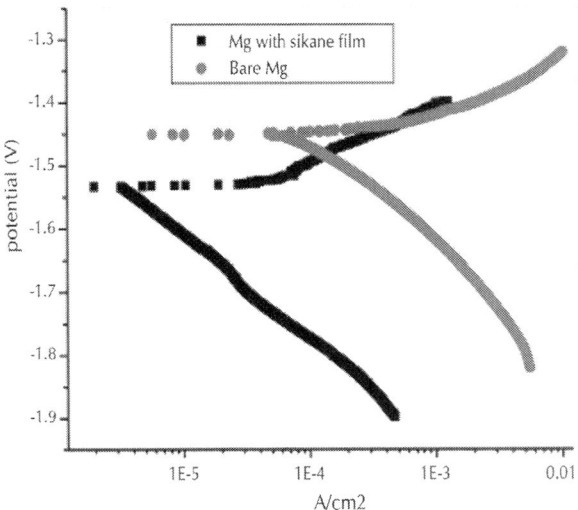

Figure 2: Polarization curve of AZ31 Mg alloy with and without silane film obtained from 8% silanic solution.

In order to investigate the influence of silane coatings on the corrosion resistance of Mg alloy, AZ31 Mg panels, treated in silane solution with different concentrations for various durations, were tested by dipping in a 5 wt.% NaCl solution. Salt spray testing was also conducted to evaluate the corrosion resistance of the silane film deposited on AZ31 Mg alloy. As indicator of the corrosion properties, the corrosion area percentages, i.e. the percentage of the surface area with corrosion pits compared to the exposed area, are shown in Fig. 3.

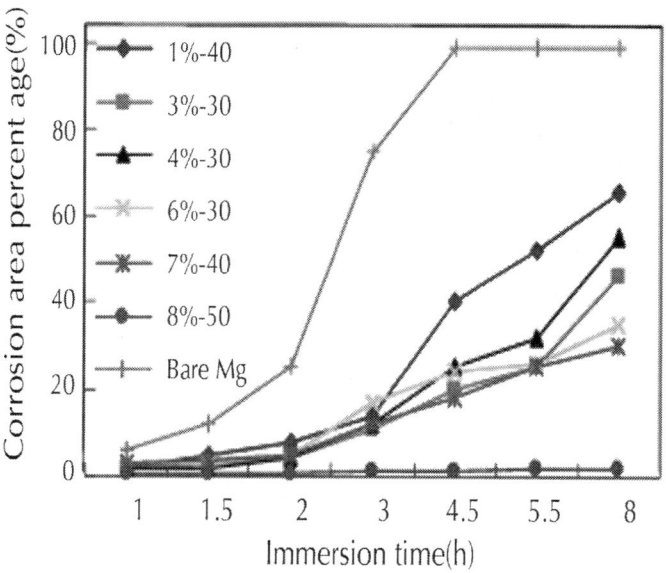

Figure 3: Corrosion area percentage of AZ31 Mg panels treated with different silane concentration from 1% to 8% and treatment time from 30 s to 50 s versus immersion time in 5 wt.% NaCl solution at room temperature.

As can be seen from Fig. 3, the corroded areas of bare Mg panels were much larger than on Mg panels with silane film after the same immersion time in NaCl solution. This indicates that Mg panels with silane film had lower corrosion rate than the bare Mg alloys. Fig. 4 shows the digital pictures of the pretreated Mg panels after salt immersion. As can be seen in Fig. 4a, the corrosion pits already appeared on bare AZ31 after the first 30 min of immersion and the corroded area fraction increased very rapidly. After 6 h the surface was totally covered with pits (Fig. 4c) and the whole panel was evenly corroded with layers of magnesium oxide and hydroxide (Fig. 4d) 24 h later.

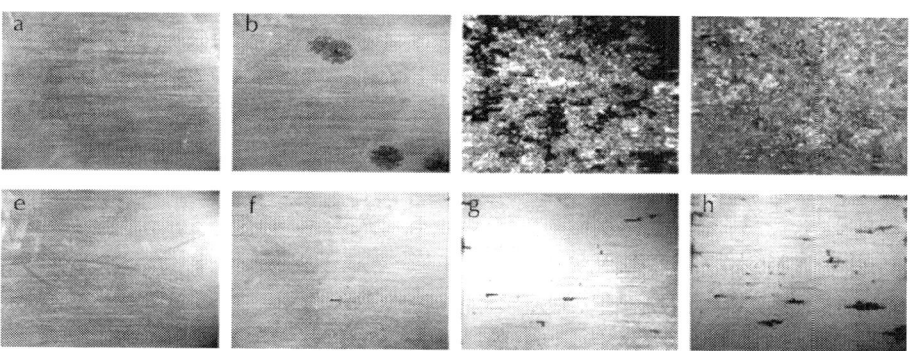

Figure 4: Digital images (a–d) bare AZ31 Mg alloy and (e–h) 8% silane pretreated AZ31 Mg alloy after dipping in 5 wt.% NaCl solution with duration of (a, e) 0 h immersion; (b, f) 0.5 h immersion; (c, g) 6 h immersion; and (d, h) 24 h immersion.

The following reactions occur when Mg substrates are immersed in a neutral NaCl solution [24]:

- Cathodic reaction:

$$O_2 + 2H_2O + 4e^- \rightarrow 4OH^-$$

(1)

$$2H_2O + 2e^- \rightarrow 2OH^- + H_2 \uparrow$$

(2)

- Anodic reaction:

$$Mg \rightarrow Mg^{2+} + 2e^-$$

(3)

- Consequently the following reaction also occurs:

$$Mg^{2+} + 2OH^- \rightarrow Mg(OH)_2 \downarrow$$

(4)

The precipitation of corrosion products onto bare Mg surfaces leads to the formation of a hydroxide layer, which thickens with immersion time and progressively hinders the corrosion processes. According to literature [19], when the corrosion rate decreases, a balance may occur between the formation and dissolution of the $Mg(OH)_2$ layer, which reaches a constant thickness. However, this layer, normally porous, is not fully protective and would allow the diffusion of charged species involved in the corrosion processes to the Mg alloy surfaces.

As shown in Fig. 4f, g, and h, some filiform-like structures appeared on the surface of Mg alloy with silane film became wider and deeper with immersion time. Normally filiform or "wormtrack" corrosion occurs under protective coatings and is caused by an active corrosion cell which moves across a metal surface [23]. The head of the filiform is the anode followed by the cathode. The tails of the filaments are covered with oxide film which is fractured by evolving H_2 [20]. The occurrence of filiform corrosion is attributed to the formation of a Si–O rich barrier film from the silanes. Additionally, the silane film modifies the chemical composition of the metallic surface. Thus, the native oxide film, which is originally a mixture of $MgO/Mg(OH)_2$, can be converted into a Mg–O–Si interface, which anchors the barrier silane film. This barrier is hydrophobic and resistant to electrolyte uptake. Thus, general corrosion did not occur on Mg alloy with silane film.

The surface corrosion morphologies examined by SEM are shown in Fig. 5. It can be seen that the corrosion products with some cracks totally covered the bare Mg surface after 24 h immersion in 5 wt.% NaCl solutions. The corrosion products, most likely a mixture of Mg oxide and Mg hydroxide, were very loose and easy to delaminate after immersion in Cl^- solution. Only few corrosion pits were found on the surface of silane pretreated Mg as shown in Fig. 5b. The typical morphology of the pits was different from the one found on bare Mg surfaces shown in Fig. 5a. The aggressive Cl^- had to penetrate the silane film before it reached the substrate and then formed a micro-cell at the pit regions. However the Mg alloy with silane film was not as severely corroded as the bare Mg alloy and the corrosion products were still adherent to the substrate (see Fig. 5c). This result was consistent with the FTIR data. The area around the corroded spots was unchanged, which indicated that the silane film had an excellent adhesion to the Mg substrate.

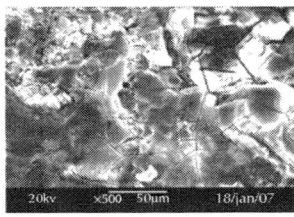 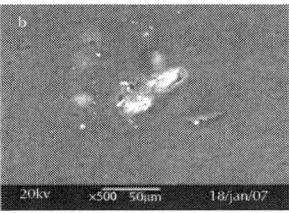 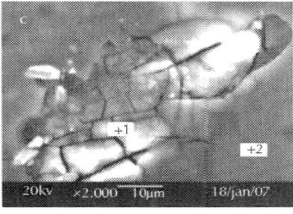

Figure 5: Corrosion surface morphology AZ31 Mg after 24 h immersion in 5 wt.% NaCl solution (a) bare metal, (b) 8% silane pretreated, and (c) the local corroded pits shown in (b) with larger magnification.

Corrosion Resistance of AZ31 Mg Alloys Silane Pretreated with E-coat

Fig. 6 shows the typical images of bare AZ31, primer painted AZ31, (AZ31/painting), bare AZ31 with E-coat (AZ31/E-coat), and silane pretreated Mg with subsequent E-coating (AZ31/Silane/E-coat) after 144 h salt spray testing. It can be seen that a great amount of corrosion product is formed on bare AZ31 panel after 144 h immersion, which is in agreement with the results reported in literature [24].

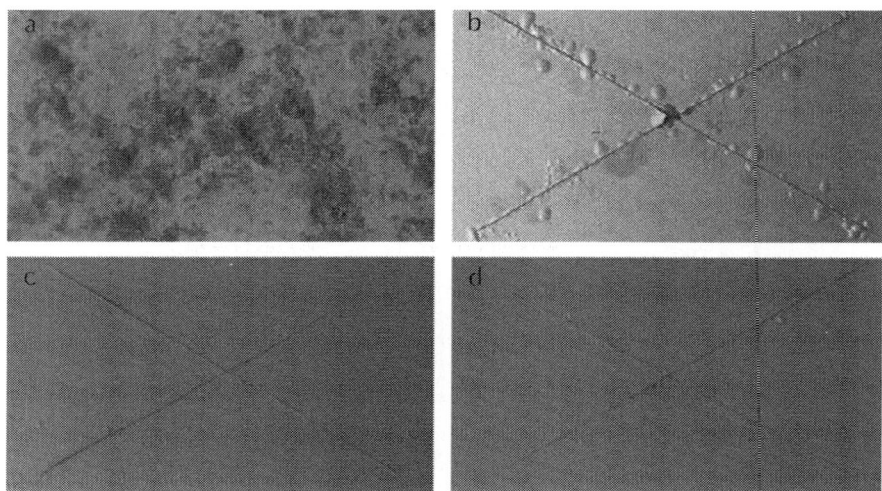

Figure 6: Images of AZ31after 144 h salt spray tested Mg panes with different treatment (a) blank, (b) paint, (c) E-coating, (d) silane/E-coating.

From Fig. 6b, it can be seen that no obvious corrosion products appeared on the surface of AZ31/painting. It should be noted that, however, there were some blisters around the scribe and the coating started to delaminate starting at the scribe. It can be concluded that the painting indeed delayed the attack of Cl species on the substrate. However, the corrosion resistance of conventional primer paint is very limited and is inferior to E-coat. In contrast, for AZ31/E-coat and AZ31/Silane/E-coat, as shown in Fig. 6c and d, no obvious blister or delamination appeared along the scribe. No corrosion products were observed on the surface of the AZ31/E-coat after 144 h immersion of salt solution. It can be seen that the panel of AZ31/Siane/E-coat had the best performance after 144 h salt spray test, with no blisters at all.

Fig. 7 shows the surface morphologies of Mg alloy panels with different coatings after 1000 h salt spray test. It can be seen that no corrosion occurred on the panel of AZ31/Silane/E-coat (Fig. 7a). However, there were several blisters on the panel of AZ31/E-coat (Fig. 7b). These blisters were formed when the corrosive species penetrated through the coating and reached the Mg surface to initiate corrosion, and thus H_2evolved as one of the corrosion products at weak areas after 1000 h of salt spray testing. The H^+ not only penetrated into the conventional paint coating shown in Fig. 7c, but also formed blisters to lift the paint coating up that eventually resulted in the coating delamination.

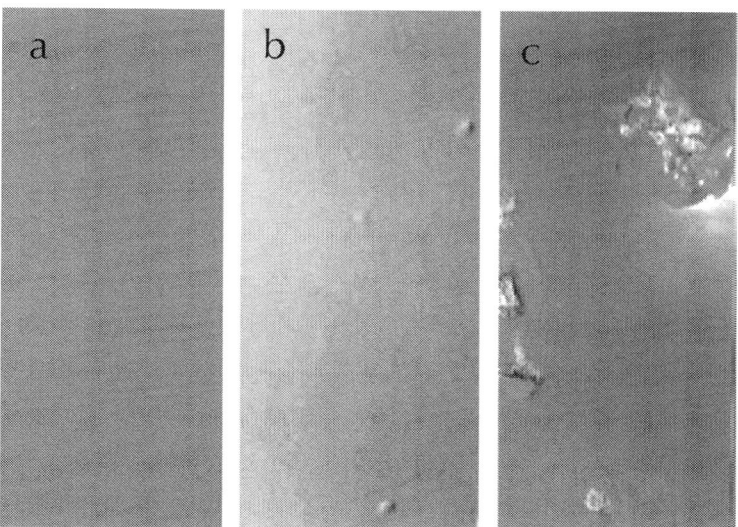

Figure 7: Images of Mg alloy with different surface treatment after 1000 h salt spray tested (a) AZ31/Silane /E-coating, (b) AZ31/E-coating, and (c) AZ31/painting.

In order to investigate their galvanic corrosion properties, AZ31 Mg alloy panels were coupled with Q235 mild steel bolts. Three groups of test panels were prepared for the dipping test in 5 wt.% NaCl solution. Typical surface images of bare Mg panels, AZ31/E-coat and AZ31/Silane/E-coat are shown in Fig. 8. As shown in Fig. 8a, serious corrosion was observed on the bare Mg panel already after 1 h immersion and also on area away from steel bolt. Only few corrosion pits were observed on the AZ31/E-coated after immersion for 4 h (Fig. 8g and k). After the same time, the bare Mg was corroded not only near the steel bolt, but the whole Mg alloy panel was damaged (Fig. 8c). On the AZ31/Silane/E-coat sample, only two to three small pits appeared around the bolt up to an immersion of 48 h. This result confirmed that the silane film has a good barrier characteristic and excellent corrosion protection to Mg substrate when compared to panels without silane pretreatment (Fig. 8e–h).

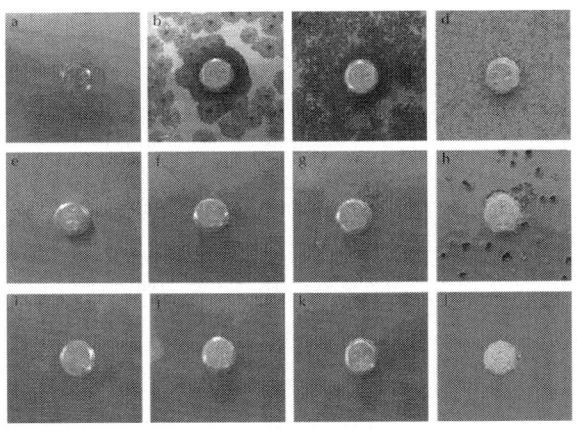

Figure 8: Digital images of AZ31 Mg alloy panels coupled with Q235 mild steel bolts tested by dipping into 5 wt.% NaCl solution. (a–d) bare Mg; (e–h) AZ31/E-coat; (i–l) AZ31/silane/E-coating after 0 h, 1 h, 4 h, 48 h immersion, respectively.

Fig. 9 shows the surface morphologies of the tested AZ31 panels after Q235 mild steel bolt after 48 h immersion in 5 wt.% NaCl solution and removal of the steel bolt. It can be seen that, on bare Mg panels, corrosion not only occurred at the location near steel bolt but also at locations away from the steel bolt. For the E-coated panels shown in Fig. 9b, several pits formed near the steel bolt, and the E-coat around the couple boundary was delaminated due to corrosion. The corrosion pits were very deep after the E-coat delaminated, as consequence of the small anode coupled with a large cathode. In contrast, as shown inFig. 9c, the panel of AZ31/ Silane/E-coat showed no obvious corrosion on the whole panel surface excepting the area underneath the steel bolt.

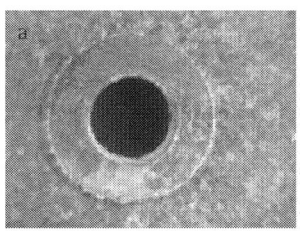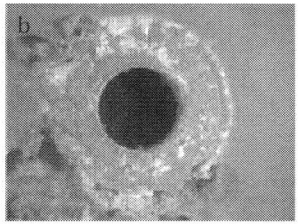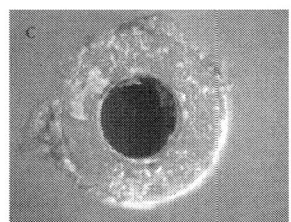

Figure 9: The morphology of AZ31 panels after the bolt of Q235 mild steel was removed with (a) bare Mg, (b) AZ31/E-coating, (c) AZ31/Silane/E-coating, exposed in 5 wt.% NaCl solution for 48 h.

It is well known that Mg alloy has a lower corrosion potential than other metals. When the steel bolt-coupled to the Mg alloy panel was exposed to salt solution, the steel bolt become the cathode and the Mg the anode of the galvanic corrosion cell. The bare Mg alloy, therefore, corroded rapidly and the whole surface area was affected due to the relative low surface resistance. On the Mg panel with coating, the Q235 bolt formed a relatively large cathode in combination with a small anode, since oxidation could only occur at defects of the Mg coating. However, for the first several hours of immersion, the coating prevented the solution from diffusion to the Mg surface and thus inhibited the initiation of corrosion. With the increase of immersion time, the corrosive species penetrated through E-coat at some areas, initiated corrosion on the Mg surfaces, and thus led the evolution of H_2 that lifted the E-coat and formed blisters (see Fig. 8h). With silane pretreatment, the thin silane film on Mg surfaces acted as an additional barrier layer to prevent corrosive species from diffusion onto AZ31 Mg alloy (Fig. 8l). However, galvanic corrosion happened at the contact area of Mg surfaces with the coupled steel bolt (Fig. 9c).

The impedance characteristic of the silane film and E-coat, the structure of the interface between Mg alloy, silane film and E-coat, and the mechanism, how the silane pretreatment improves adhesion and corrosion resistance of Mg, are still under investigation and the results will be reported in a subsequent paper.

CONCLUSIONS

The E-coats deposited on Mg alloys have excellent corrosion resistance which can be improved by silane pretreatment. Silane pretreatment of Mg alloy substrate will not only improve the adhesion of the E-coat to the Mg substrate, but delays the diffusion of corrosive species to the Mg substrate. The galvanic corrosion of AZ31 coupled with mild steel Q235 bolt was, therefore, greatly inhibited.

ACKNOWLEDGMENTS

The authors would like to thank the Project No. 9140A18060409 QT0202 for financial support. Many thanks for the experimental assistant from Chongqing Institute of Technology and Institute of Southwest Engineering Technology.

REFERENCES

1. J.E. Hillis, M. Pekguleryuz, I. Nakatsugawa, ASM International, Materials Park, OH, 1999, pp. 194–210.

2. G.R. Pilcher, Macromol. Sympos. 1 (2002) 187.

3. W.J. Van Ooij, United States, United States Patent, 5,108,793, (1992) (28 April).

4. V. Palanivel, Y. Huang, W.J. van Ooij, Prog. Org. Coat. 53 (2) (2005) 153–168.

5. V. Palanivel, D. Zhu, W.J. Van Ooij, Prog. Org. Coat. (2003) 47–384.

6. F. Deflorian, S. Rossi, L. Fedrizzi, Electrochim. Acta 51 (27) (2006) 6097–6103.

7. M.F. Montemon, M.G.S. Ferreira, Electrochim. Acta 52 (2007) 7486–7495.

8. A. Bakkar, V. Neubert, Corros. Sci. 49 (2007) 1110.

9. J.E. Gray, B. Luan, J. Alloys Compd. 336 (2002) 88.

10. J.H. Nordlien, S. Ono, N. Masuko, K. Nisancioglu, Corros. Sci. 39 (1997) 1397.

11. Wieliczka, David, Deffeyes et al., United States, United States Patent, 5,312,529, (1994) (17 May).

12. C.E. Moffitt, C.M. Reddy, Q.S. Yu, D.M. Wieliczka, Corrosion 56 (10) (2000) 1032–1045.

13. J. Zhang, Y. Chan, Q. Yu, Prog. Org. Coat. 61 (2008) 28–37.

14. Zhang Jin, Sun Zhifu, Chinese, Chinese Patent, 200510057166.8, 2005.7.

15. C.E. Barchiche, E. Rocca, C. Juers, J. Hazan, Electrochim. Acta 53 (2007) 417–425.

16. Y. Zhang, C. Yan, Surf. Coat. Technol. 201 (2006) 2381–2386.

17. Z. Jin, W. Chaoyun, H. Fuxiang, Z. Wei, J. Chin. Soc. Corros. Protect. 28 (3) (2008) 146–150.

18. M.F. Montemor, M.G.S. Ferreira, Electrochim. Acta 52 (2007) 7486–7495.

19. Danqing, Zhu, Ph.D. Dissertation, University of Cincinnati, USA (2005).

20. X. Yu, G. Li, J. Alloys Compd. 364 (2004) 193.

21. A.J. Davenport, H.S. Isaacs, M.W. Kendig, Corros. Sci. 32 (1992) 653.

22. G. Baril, N. Pébére, Corros. Sci. 43 (2001) 471.

23. G. Song, A. Atrens, Adv. Eng. Mater. 1 (1999) 1438–1656.

24. A. Pardoa, M.C. Merino, A.E. Coy, et al., Electrochim. Acta 53 (27) (2008) 7890–7902.

Influence of Conductivity on Cathodic Protection of Reinforced Alkali-Activated Slag Mortar Using the Finite Element Method

R. Montoya[a], W. Aperador[b], and D.M. Bastidas[c]

[a]Mathematics Department, Chemistry Faculty, National Autonomous University of Mexico, UNAM, University City, 04510 Mexico, DF, Mexico
[b]Composite Materials Group, Universidad del Valle, Cali, Colombia
[c]CENIM-National Centre for Metallurgical Research, CSIC, Avda. Gregorio del Amo 8, 28040 Madrid, Spain

ABSTRACT

Cathodic protection (CP) is considered to be the only rehabilitation method for chloride-induced rebar corrosion in reinforced concrete structures. The protection current distribution depends on several parameters, such as the geometry and number of rebars and the concrete resistivity. In order to investigate the influence of concrete resistivity on the possibilities and limitations of rebar protection, this paper presents a numerical approach based on the finite element method (FEM) in conjunction with laboratory results to determine its impact on the CP of alkali-activated slag mortar. An ordinary Portland cement was also tested for comparative purposes.

INTRODUCTION

Cathodic protection (CP) as a rehabilitation method has proven to stop corrosion in salt-contaminated bridge decks, regardless of whether or not the concrete contains chlorides [1]. CP extends the service life of buried steel pipelines, oil and gas well casings, offshore oil-drilling structures, seagoing ship hulls, marine piles, water tanks and chemical equipment. The concept behind CP consists of shifting the electrode potential of a metal to a more negative value where the corrosion rate is sufficiently low to suppress the anodic reaction [2]:

$$Fe \rightarrow Fe^{2+} + 2e^-$$

(1)

and the cathodic reaction is enhanced:

$$O_2 + 2H_2O + 4e^- \rightarrow 4OH^-$$

(2)

$$2H^+ + 2e^- \rightarrow H_2$$

(3)

thus decreasing the overall corrosion current. Oxygen reduction (Eq. (2)) is the main cathodic reaction in concrete, because concrete has a high pH and oxygen is thermodynamically a far more powerful electron acceptor than the hydrogen ion (Eq. (3)). The direct current (DC) for CP systems can be supplied either via mains power in impressed current CP systems (ICCP) or by a sacrificial anode CP system (SACP). In a SACP device, single or multiple anodes distribute the cathodic current to the protected structure. For buried structures the anodes are often inert graphite. For immersed seawater structures they may be high-silicon cast iron or platinum-coated titanium. Magnesium, zinc, aluminium, and aluminium–zinc–indium alloy sacrificial anodes, welded to buried and immersed structures, provide long-term CP.

The very high temperatures (1400–1500 °C) required to manufacture ordinary Portland cement (OPC) account for the high cost of this process, which is responsible for 40% of all energy consumed. Indeed, the cement industry is regarded to generate 6–7% of all greenhouse gases emitted world-wide [3]. Thus, granulated blast furnace slag (GBFS) may represent an option for emission mitigation. An alkali-activated cement is a system in which an alkaline activator promotes the pozzolanic reactions on an inorganic solid of natural or artificial origin to generate a material with cementitious characteristics. One such material is alkali-activated slag (AAS) cement, which is the result of mixing GBFS and alkaline substances [4].

The aim of this paper is to assess the effectiveness of CP of reinforced AAS and OPC mortars with different conductivities using a numerical approach based on the finite element method (FEM). A two-dimensional simulation of the potential distribution was developed and its accuracy was verified by laboratory results using the criterion of measured polarized potential values between −0.85 and −1.0 V vs. a copper/copper sulphate electrode (Cu/CuSO$_4$) (CSE) [5].

NUMERICAL METHOD

Numerical methods, such as the finite difference method (FDM), boundary element method (BEM) and FEM, have been used to design CP and electrochemical systems [6]. These methods have been adapted to calculate the current and potential distributions in complex structures. FDM was one of the first methods developed and presents the disadvantage of its instability in complex geometry systems. This limitation is partially solved using the FEM technique. Numerical methods require the discretization of a large volume domain, generating a great number of equations, but may be simply applied in the case of inhomogeneous domains. They were also initially used to predict steel corrosion behaviour addressing coplanar electrode configurations with varying degrees of polarization effects, generally limited to a one-dimensional approach.

In all of these numerical treatments the governing field equations are the Poisson and Laplace equations:

$$-\kappa\Delta\phi = -\kappa\left(\frac{\partial^2\phi}{\partial x^2} + \frac{\partial^2\phi}{\partial y^2} + \frac{\partial^2\phi}{\partial z^2}\right) = f(x,y,z)$$

(4)

with the additional assumption that the electrolytic medium and electrode materials possessed constant electrical properties, where, in Eq. (4), f(x,y,z)=0 for Laplace's equation; f(x,y,z)≠0 for Poisson's equation [7]; is Laplace's operator; κ is conductivity; and φ is the electrical potential. The BEM and FEM techniques have been used for numerical modelling of galvanic corrosion and CP [8], [9], [10], [11],[12], [13] and [14], and improved using the penalty function method (PFM) and a multipeaked cost function (MCF), which was coupled with a genetic algorithm method (GAM) combined with the conjugate gradient method (CGM) [15]. Pipelines have also been CP modelled by coupling the FEM and BEM techniques and including non-negligible ohmic voltage drops (IR) [16]. CP of coated pipelines with bare metal surfaces has been modelled using the BEM technique coupled with an iterative Newton–Raphson algorithm [17] and [18].

The Two-Dimensional Boundary Value Problem

Considering the physical situation shown in Fig. 1a, a two-dimensional square container with a reinforced cylindrical mortar specimen and an AISI 304 stainless steel (SS) anode has been developed. The corresponding boundary value problem is:

$$-\kappa\Delta\phi = f(x, y)$$

(5)

with (x, y) coordinates in the Ω domain, which is the union of Ω_1 and Ω_2 domains (see Fig. 1a), i.e. the steel/mortar system within one domain (Ω_1) and the electrolyte (NaCl) within the other domain (Ω_2), where κ and ϕ have been defined above; and f(x,y) is the ICCP externally applied to the steel rebar/mortar system through the AISI 304 SS anode.

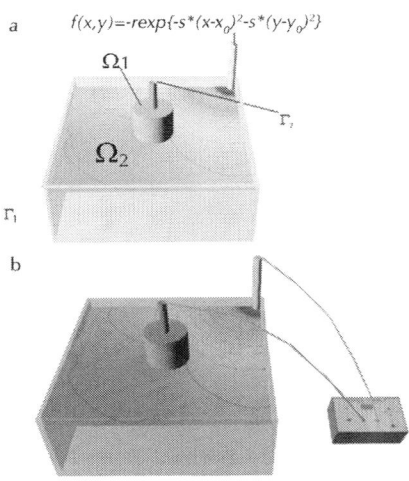

Figure 1: (a) Schematic representation of the experimental design used for cathodic protection (CP) numerical simulation. $_{\Gamma1}$ ($-K\dfrac{\partial\phi}{\partial n} = 0$) and $-K\dfrac{\partial\phi}{\partial n} = h(\phi)$ $_{\Gamma2}$ ( ) are the boundaries of the structural 1018 steel rebar and square plastic container, respectively. The Ω_1 and Ω_2 domains refer to the NaCl conductivity and cylindrical reinforced mortar conductivity, respectively. (b) Scheme with the container/reinforced mortar/AISI 304 SS anode system and the externally impressed current cathodic protection (ICCP) source.

The boundary conditions are: $-K\dfrac{\partial \phi}{\partial n} = 0$ in Γ_1 (container) and

$-K\dfrac{\partial \phi}{\partial n} = h(\phi)$ in Γ_2 (rebar) (see Fig. 1a), where n is the external vector normal to the boundary; and h(ϕ) is the function that represents the polarization curve of the steel rebar immersed in 3.5% NaCl electrolyte. Its variational formulation was carried out in the same way as in Ref. [13], but in the present study the polarization curve was used directly, instead of the Butler–Volmer equation, in Γ_2 (rebar) (see Fig. 1a). The Ω domain was discretized into 925 elements and the Gauss–Seidel method was employed to solve the system thus established. The mathematical function used to approximate the AISI 304 SS anode was:

$$f(x, y) = -r\exp\left\{-s^*(x - x_0)^2 - s^*(y - y_0)^2\right\}$$

(6)

where f(x,y) is the externally applied ICCP (as indicated above); r is a factor involving the impressed current applied; s^* is a factor proportional to the diameter of the AISI 304 SS anode; and x_0 and y_0 are the rectangular coordinates of the AISI 304 SS electrode (see Fig. 1a).

EXPERIMENTAL

GBFS from the company "Acerías Paz del Río", located in Boyacá, Colombia, was used with a chemical composition (% by weight) of 33.7 SiO_2, 12.8 Al_2O_3, 3.09 Fe_2O_3, 45.4 CaO, 0.5 TiO_2, 0.64 SO_3, 1.79 MgO, and 2.08 ignition loss; a specific surface area of 398 m^2 kg^{-1}; and a specific gravity of 2860 kg m^{-3}. The basicity index (CaO + MgO)/(SiO_2 + Al_2O_3) and quality index (CaO + MgO + Al_2O_3)/(SiO_2 + TiO_2) were 1.01 and 1.73, respectively. According to ASTM C 989-99 this is grade 80 slag [19]. The waterglass used as the activating solution consisted of a mix of commercial sodium silicate

(31.7% SiO_2, 12.3% Na_2O, and 56.0% water) and a 50% NaOH solution to obtain a SiO_2/Na_2O ratio of 2.4. The Na_2O concentration in the waterglass activating solution added to the mortar was 5% by weight of slag. The aggregates used were a siliceous gravel with a maximum grain size of 19 mm, specific gravity of 2940 kg m^{-3}, and 1.3% absorption. OPC according to ASTM C 150-02 was also tested for comparative purposes [20], with a specific gravity of 2990 kg m^{-3} and a specific surface area of 400 kg m^{-3}. AAS and OPC mortars were prepared with a water/cement ratio of 0.4. The AAS and OPC specimens were de-moulded after 24 h and cured in a climatic cabinet for 28 days, at 90% relative humidity (RH) for AAS specimens and at 100% RH for OPC specimens to prevent leaching of the activating solutions and to assure that the hydration reaction and product formation processes were not affected.

Structural 1018 steel rebars of 6.35 mm diameter were used according to ASTM A 706-08 [21]. Reinforced cylindrical mortar specimens (10 cm length and 5 cm diameter) were used to perform the experimental tests. An active surface area of 10 cm^2 was marked on the working electrode with an epoxy resin, thus isolating the triple mortar/carbon steel/atmosphere interface to avoid possible localized corrosion attack due to differential aeration.

Carbonation exposure was performed by accelerated testing in a cabinet with 3% CO_2, 65% RH, and 25 °C (mortars AASA and OPCA) and exposure in a laboratory environment (Universidad del Valle, Cali, Colombia) with 0.03% CO_2, 65% RH, and 25 °C (mortars AASL and OPCL). Polarization measurements were performed up to 49 days by immersing the reinforced mortar specimens in a 3.5% NaCl electrolyte using a CSE reference electrode. Fig. 1a shows the device used to perform CP, consisting of a 13 × 13 cm two-dimensional square plastic container with a reinforced cylindrical mortar specimen in the centre and an AISI 304 SS anode of 9 mm diameter situated at the point at $x_0 = 12.5$ cm and $y_0 = 12.5$ cm. Fig. 1b shows a scheme with the container/reinforced mortar/AISI 304 SS anode system and the externally ICCP source used for CP application. Table 1 lists the parameters used for the numerical simulation.

Table 1: Impressed current cathodic protection (ICCP) design parameters for reinforced alkali-activated slag mortar in accelerated carbonation conditions (AASA), reinforced ordinary Portland cement mortar in accelerated carbonation conditions (OPCA), reinforced alkali-activated slag mortar in laboratory conditions (AASL), and reinforced ordinary Portland cement mortar in laboratory conditions (OPCL)

Parameter	Description
AASA mortar	
Average conductivity of AASA mortar	0.019 mS cm^{-1}
Average conductivity of 3.5% NaCl electrolyte	86.3 mS cm^{-1}
Anode axis location (x_0, y_0)	$x_0 = 12.5$ cm; $y_0 = 12.5$ cm
Impressed current cathodic protection (ICCP)	55×10^{-6} A cm^{-2}
OPCA mortar	
Average conductivity of OPCA mortar	0.042 mS cm^{-1}
Average conductivity of 3.5% NaCl electrolyte	86.3 mS cm^{-1}
Anode axis location (x_0, y_0)	$x_0 = 12.5$ cm; $y_0 = 12.5$ cm
Impressed current cathodic protection (ICCP)	55×10^{-6} A cm^{-2}
AASL mortar	
Average conductivity of AASL mortar	0.086 mS cm^{-1}
Average conductivity of 3.5% NaCl electrolyte	86.3 mS cm^{-1}
Anode axis location (x_0, y_0)	$x_0 = 12.5$ cm; $y_0 = 12.5$ cm
Impressed current cathodic protection (ICCP)	55×10^{-6} A cm^{-2}
OPCL mortar	
Average conductivity of OPCL mortar	0.23 mS cm^{-1}
Average conductivity of 3.5% NaCl electrolyte	86.3 mS cm^{-1}
Anode axis location (x_0, y_0)	$x_0 = 12.5$ cm; $y_0 = 12.5$ cm
Impressed current cathodic protection (ICCP)	55×10^{-6} A cm^{-2}

RESULTS AND DISCUSSION

Fig. 2, Fig. 3, Fig. 4 and Fig. 5 show the calculated isopotential line and potential distributions (top views inFigs. 2a, 3a, 4a, and 5a and side views in Figs. 2b, 3b, 4b, and 5b) for reinforced AASA (Fig. 2), OPCA (Fig. 3), AASL (Fig. 4), and OPCL (Fig. 5) using a 2-dimensional numerical simulation.

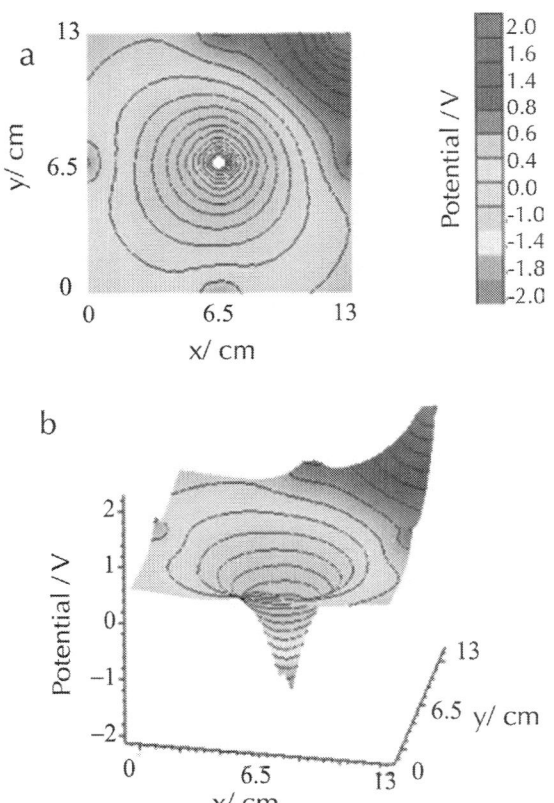

Figure 2: (a) Calculated isopotential lines distribution using rectangular (x_0, y_0) coordinates for reinforced AASA mortar. The polarized potential variation between units was 0.2 V. (b) A 3-D view of the calculated potential distribution for reinforced AASA mortar. The anode was in the position of $x_0 = 12.5$ cm, $y_0 = 12.5$ cm.

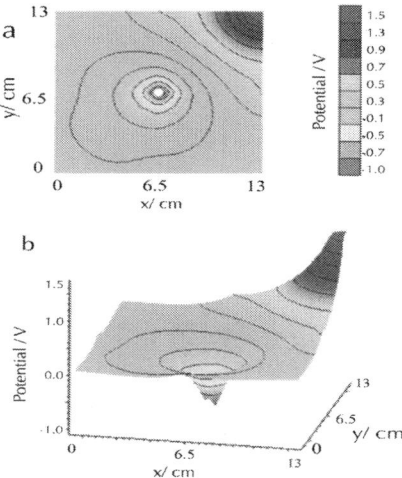

Figure 3: (a) Calculated isopotential lines distribution using rectangular (x_0, y_0) coordinates for reinforced OPCA mortar. The polarized potential variation between units was 0.2 V. (b) A 3-D view of the calculated potential distribution for reinforced OPCA mortar. The anode was in the position of $x_0 = 12.5$ cm, $y_0 = 12.5$ cm.

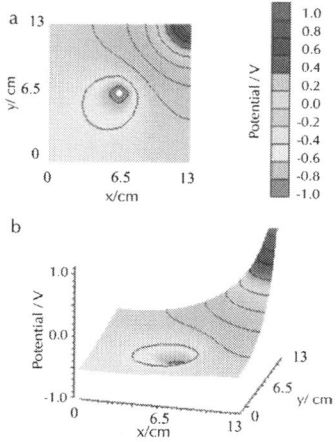

Figure 4: (a) Calculated isopotential lines distribution using rectangular

(x_0, y_0) coordinates for reinforced AASL mortar. The polarized potential variation between units was 0.2 V. (b) A 3-D view of the calculated potential distribution for reinforced AASL mortar. The anode was in the position of x_0 = 12.5 cm, y_0 = 12.5 cm.

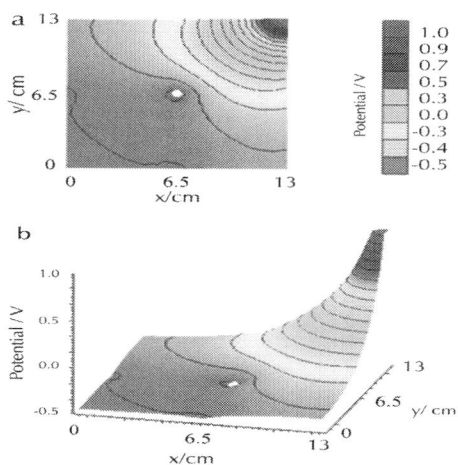

Figure 5: (a) Calculated isopotential lines distribution using rectangular (x_0, y_0) coordinates for reinforced OPCL mortar. The polarized potential variation between units was 0.1 V. (b) A 3-D view of the calculated potential distribution for reinforced OPCL mortar. The anode was in the position of x_0 = 12.5 cm, y_0 = 12.5 cm.

The potential variation between isopotential lines was 0.1 V vs. CSE for OPCL (Fig. 5a) and 0.2 V vs. CSE for AASA (Fig. 2a), OPCA (Fig. 3a) and AASL (Fig. 4a).

The white circle (in the middle position) in Figs. 2a, 3a, 4a, and 5a indicates the position of the rebar, and the cone with the red vertex in Figs. 2b, 3b, 4b, and 5b (initiation of cone in the latter) indicates the potential value of the rebar.

Table 2 summarizes the rebar protection potential obtained using the 2-dimensional numerical simulation, and Table 3 includes the laboratory results of the rebar protection potential measured after 49 days experimentation.

Table 2: Rebar protection electrical potential obtained using the numerical simulation

Mortar	Potential, V vs. CSE
AASA	−1.980
OPCA	−0.970
AASL	−0.980
OPCL	−0.510

Table 3: Laboratory results of the rebar protection electrical potential measured after 49 days experimentation

Mortar	Potential, V vs. CSE
AASA	−0.892
OPCA	−0.943
AASL	−0.952
OPCL	−0.897

It was assumed that the 1018 steel rebar was cathodically protected according to the criterion of the measurement of polarized potential values between −0.85 and −1.0 V vs. CSE [5]. The quantitative results of Fig. 2 for AASA mortar show that the 55×10^{-6} A cm^{-2} externally applied ICCP originated an electrical potential of −1.980 V vs. CSE (see Table 2). Nevertheless, an overprotection was reached which was visualized by the red colour in Fig. 2. So the steel rebar may be cathodically protected using a lower external ICCP than the 55×10^{-6} A cm^{-2} used.

In contrast to AASA, which has the lowest conductivity (0.019 mS cm^{-1}), OPCL with the highest conductivity (0.23 mS cm^{-1}) and whose quantitative results are shown in Fig. 5, represents the opposite case. On OPCL the externally applied ICCP, 55×10^{-6} A cm^{-2}, caused a shift in the electrical potential to −0.510 V vs. CSE (see Table 2), which means that the externally applied ICCP was not enough to protect the specimen and the steel rebar was actively corroding, as visualized by the red colour in Fig. 5. Thus, one order of conductivity, 0.019 mS cm^{-1} for AASA and 0.23 mS cm^{-1} for OPCL, caused a difference of more than 1.4 V in the electrical potential. In the latter example (Fig. 5) the ICCP required to protect the steel rebar was higher than the 55 × 10^{-6} A cm^{-2} utilized. Comparing the results of Fig. 2 with Fig. 5, the use of AASA was more attractive than OPCL from a CP point of view because the former required a lower ICCP than the latter.

Fig. 3 for OPCA and Fig. 4 for AASL show an intermediate situation between AASA (Fig. 2) and OPCL (Fig. 5). On OPCA and AASL the externally applied ICCP (55 × 10^{-6} A cm^{-2}) originated an electrical potential of approximately −1.0 V vs. CSE (see Table 2), which means that the CP of the steel rebar was effective, according to the criterion of measured polarized potential values between −0.85 and −1.0 V vs. SCE.

Therefore, conductivity differences of one order of magnitude between AASA (0.019 mS cm^{-1}) and OPCL (0.23 mS cm^{-1}) drastically influence the CP design of the steel/mortar system. OPCA (conductivity 0.042 mS cm^{-1}) and AASL (conductivity 0.086 mS cm^{-1}) were well cathodically protected (see Figs. 3 and 4). These results are of practical importance because they are a good example of using CP engineering to save energy when protecting reinforced concrete structures. Numerical simulations based on FEM calculations allow CP performance to be assessed using conductivity as a variable in the numerical simulation. It should be noted that the   parameter (conductivity) that appears in Eq. (5) considers the use of Ω_1 and Ω_2 domains to model the CP of the steel rebar/mortar system using the FEM technique, which shows the great flexibility of this method in the Ω domain conditions.

Comparing the rebar protection electrical potential calculated using the numerical simulation (Table 2) with laboratory results of the rebar protection electrical potential measured after 49 days experimentation (Table 3), it may be indicated that the numerical simulation approach utilised in the present paper is a good tool for reinforced mortar CP design. Nevertheless, the discrepancy between the protection electrical potential yielded using the numerical method and the electrical potential measured in the laboratory results for AASA and OPCL (with the lowest and highest conductivity, respectively) may be attributed to the "ideal" situation simulated numerically, i.e. the number of rebars, their geometry, and the corroding and passive zones, etc. The presence of pores and irregularities in the mortar was not considered.

CONCLUSIONS

AASA mortar presents the best conductivity properties for CP engineering design. OPCL, with a conductivity that is one order of magnitude higher than AASA, requires a high externally impressed current cathodic protection. The order of the mortars in terms of the required impressed current density, from low to high, was: AASA < OPCA < AASL < OPCL. This result is of practical importance because it is a good example of using CP engineering to save energy when protecting reinforced concrete structures, and as expected corroborates Ohm's law. Numerical methods based on FEM calculation allow CP performance to be assessed using the conductivity parameter. From a CP point of view, the carbonation process has a positive influence on the conductivity parameter, favouring CP design.

The accuracy of the proposed numerical results was assessed by comparisons with laboratory results measured after 49 days experimentation. The simulation approach was a good procedure for the design of AASL and OPCA CP with intermediate conductivity values of the order of 0.06 mS cm^{-1}. However, the discrepancy between the numerical method and the laboratory results for AASA and OPCL, with low (0.02 mS cm^{-1}) and high (0.23 mS cm^{-1}) conductivities, respectively, may be attributed to the "ideal" and simple situation simulated, considering conductivity as the only variable.

ACKNOWLEDGMENTS

R. Montoya expresses his gratitude to the Subdivision of Academic Formation and the DGAPA of UNAM, Mexico, and CSIC, Spain, for the scholarship granted to him. W. Aperador expresses his gratitude to the Centre of Excellence in Novel Materials (CENM) and COLCIENCIAS of Colombia, Project Geoconcret, for the scholarship granted to him. The authors express their gratitude to Project BIA2008-05398 from CICYT, Spain, for financial support.

REFERENCES

1. B.S. Wyatt, D.J. Irvine, A review of cathodic protection of reinforced-concrete, Mater. Perform. 26 (1987) 12–21.

2. J.H. Morgan, Cathodic Protection. Its Theory and Practice in the Prevention of Corrosion, Barnicotts, London, 1959, p. 3.

3. D.M. Bastidas, A. Fernández-Jiménez, A. Palomo, J.A. González, A study on the passive state stability of steel embedded in activated fly ash mortars, Corros. Sci. 50 (2008) 1058–1065.

4. E. Rodríguez, S. Bernal, R. Mejía de Gutiérrez, F. Puertas, Alternative concrete based on alkali-activated slag, Mater. Constr. 58 (2008) 53–67.

5. M. Funahashi, J.B. Bushman, Technical review of 100 mV polarization shift criterion for reinforcing steel in concrete, Corrosion 47 (1991) 376–386.

6. R.D. Strommen, R.S. Munn (Eds.), Computer Modelling in Corrosion, ASTM STP 1154, Philadelphia, PA, USA, 1992, pp. 229–447.

7. R.G. Kasper, M.G. April, Electrogalvanic finite-element analysis of partially protected marine structures, Corrosion 39 (1983) 181–188.

8. R.S. Munn, O.F. Devereux, Numerical modelling and solution

of galvanic corrosion systems 1. Governing differential-equation and electrodic boundaryconditions, Corrosion 47 (1991) 612–618.

9.   J.X. Jia, G. Song, A. Atrens, D.S. John, J. Baynham, G. Chandler, Evaluation of the BEASY program using linear and piecewise linear approaches for the boundary conditions, Mater. Corros. 55 (2004) 845–852.

10. A. Canelas, J. Herskovits, J.C.F. Telles, Shape optimization using the boundary element method and a SAND interior point algorithm for constrained optimization, Comput. Struct. 86 (2008) 1517–1526.

11. V.G. DeGiorgi, S.A. Wimmer, Geometric details and modelling accuracy requirements for shipboard impressed current cathodic protection system modelling, Eng. Anal. Boundary Elem. 29 (2005) 15–28.

12. S.H. Lee, D.W. Townley, K.O. Eshun, A boundary element model of cathodic well casing protection, J. Comput. Phys. 107 (1993) 338–347.

13. R. Montoya, O. Rendón, J. Genesca, Mathematical simulation of cathodic protection system by finite element method, Mater. Corros. 56 (2005) 404– 411.

14. P. Lambert, P.S. Mangat, F.J. O'Flaherty, Y.-Y. Wu, Influence of resistivity on current and potential distribution of cathodic protection systems for steel framed masonry structures, Corros. Eng. Sci. Technol. 43 (2008) 16–22.

15. K. Amaya, S. Aoki, Effective boundary element methods in corrosion analysis, Eng. Anal. Boundary Elem. 27 (2003) 507–519.

16. F. Brichau, J. Deconinck, A numerical-model for cathodic protection of buried pies, Corrosion 50 (1994) 39–49.

17. M.E. Orazem, J.M. Esteban, K.J. Kennelley, R.M. Degerstedt, Mathematical models for cathodic protection of an underground pipeline with coating holidays: part 1. Theoretical development, Corrosion 53 (1997) 264–272.

18. M.E. Orazem, J.M. Esteban, K.J. Kennelley, R.M. Degerstedt,

Mathematical models for cathodic protection of an underground pipeline with coatingholidays: part 2. Case studies of parallel anode cathodic protection systems, Corrosion 53 (1997) 427–436.

19. ASTM C 989-99 Standard, Standard specification for ground granulated blastfurnace slag for use in concrete and mortars, American Society for Testing and Materials, West Conshohocken, PA, 1999.

20. ASTM C 150-02 Standard, Standard specification for Portland cement, American Society for Testing and Materials, West Conshohocken, PA, 2002.

21. ASTM A 706-08 Standard, Standard specification for low-alloy steel deformed and plain bars for concrete reinforcement, American Society for Testing and Materials, West Conshohocken, PA, 2008.

Citations

CHAPTER 1

Hikmet Altun, Hakan Sinici, Corrosion behaviour of magnesium alloys coated with TiN by cathodic arc deposition in NaCl and Na2SO4 solutions, Materials Characterization, Volume 59, Issue 3, March 2008, Pages 266-270, ISSN 1044-5803, http://dx.doi.org/10.1016/j.matchar.2007.01.004.

CHAPTER 2

B. E. Amitha Rani and Bharathi Bai J. Basu, "Green Inhibitors for Corrosion Protection of Metals and Alloys: An Overview,"

International Journal of Corrosion, vol. 2012, Article ID 380217, 15 pages, 2012. doi:10.1155/2012/380217.

CHAPTER 3

Belén Díaz, Jolanta wiatowska, Vincent Maurice, Antoine Seyeux, Emma Härkönen, Mikko Ritala, Sanna Tervakangas, Jukka Kolehmainen, Philippe Marcus, Tantalum oxide nanocoatings prepared by atomic layer and filtered cathodic arc deposition for corrosion protection of steel: Comparative surface and electrochemical analysis, Electrochimica Acta, Volume 90, 15 February 2013, Pages 232-245, ISSN 0013-4686, http://dx.doi.org/10.1016/j.electacta.2012.12.007.

CHAPTER 4

El-Sayed M. Sherif and Abdulhakim A. Almajid, "Anodic Dissolution of API X70 Pipeline Steel in Arabian Gulf Seawater after Different Exposure Intervals," Journal of Chemistry, vol. 2014, Article ID 753041, 7 pages, 2014. doi:10.1155/2014/753041.

CHAPTER 5

Zhengfeng Li, Fuxing Gan, Xuhui Mao, A study on cathodic protection against crevice corrosion in dilute NaCl solutions, Corrosion Science, Volume 44, Issue 4, April 2002, Pages 689-701, ISSN 0010-938X, http://dx.doi.org/10.1016/S0010-938X(01)00042-7.

CHAPTER 6

Saeed Mohammadi, Fatemeh Baghaei Ravari, and Athareh Dadgarinezhad, "Improvement in Corrosion Inhibition Efficiency of Molybdate-Based Inhibitors via Addition of Nitroethane and Zinc in Stimulated Cooling Water," ISRN Corrosion, vol. 2012, Article ID 515326, 9 pages, 2012. doi:10.5402/2012/515326.

CHAPTER 7

Emma Härkönen, Sanna Tervakangas, Jukka Kolehmainen, Belén Díaz, Jolanta wiatowska, Vincent Maurice, Antoine Seyeux, Philippe Marcus, Martin Fenker, Lajos Tóth, György Radnóczi, Mikko Ritala, Interface control of atomic layer deposited oxide coatings by filtered cathodic arc deposited sublayers for improved corrosion protection, Materials Chemistry and Physics, Volume 147, Issue 3, 15 October 2014, Pages 895-907, ISSN 0254-0584, http://dx.doi.org/10.1016/j.matchemphys.2014.06.035.

CHAPTER 8

Jin Zhang, Chaoyun Wu, Corrosion protection behavior of AZ31 magnesium alloy with cathodic electrophoretic coating pretreated by silane, Progress in Organic Coatings, Volume 66, Issue 4, December 2009, Pages 387-392, ISSN 0300-9440, http://dx.doi.org/10.1016/j.porgcoat.2009.09.001.

CHAPTER 9

R. Montoya, W. Aperador, D.M. Bastidas, Influence of conductivity on cathodic protection of reinforced alkali-activated slag mortar using the finite element method, Corrosion Science, Volume 51, Issue 12, December 2009, Pages 2857-2862, ISSN 0010-938X, http://dx.doi.org/10.1016/j.corsci.2009.08.020.

Index